I0033034

Alexander Ritchie Leask

Refrigerating Machinery

Its principles and management. With sixty-four illustrations.

Alexander Ritchie Leask

Refrigerating Machinery
Its principles and management. With sixty-four illustrations.

ISBN/EAN: 9783337163945

Printed in Europe, USA, Canada, Australia, Japan

Cover: Foto ©berggeist007 / pixelio.de

More available books at **www.hansebooks.com**

REFRIGERATING
MACHINERY

ITS PRINCIPLES AND MANAGEMENT

WITH SIXTY-FOUR ILLUSTRATIONS

By A. RITCHIE LEASK

AUTHOR OF

"TRIPLE AND QUADRUPLE EXPANSION ENGINES AND BOILERS AND
THEIR MANAGEMENT"

"BREAKDOWNS AT SEA AND HOW TO REPAIR THEM"

LONDON

TOWER PUBLISHING COMPANY LD.

95, MINORIES, E.

1895

PREFACE

—◆—

OWING to the great increase in the number and importance of refrigerating plants, both afloat and ashore, a
demand has arisen for a work dealing more or less fully
with this subject. This the Author has endeavoured to
produce, with what measure of success remains to be seen.

For permission to reproduce some of the illustrations
and tables the Author is indebted to the Council of the
Institution of Mechanical Engineers, and to the Frick
Co. of America, the Linde British Refrigeration Co.,
Messrs. Sterne & Co., Messrs. J. & E. Hall Limited, and
the Pulsometer Engineering Co., for their courtesy in
furnishing illustrations and particulars of their various
specialties.

The Author also takes this opportunity of expressing
his indebtedness to Mr. T. B. Lightfoot for generously
according permission to use a number of papers read
by him before the various societies, and for assistance
rendered in other ways.

CONTENTS

CHAPTER I

CHAPTER II

CHAPTER III

CHAPTER IV

CHAPTER V

CHAPTER VI

Contents

CHAPTER VII

LIST OF ILLUSTRATIONS

⠄⠄⠄⠄⠄⠄⠄⠄
⠄⠄⠄⠄⠄⠄⠄⠄

REFRIGERATING MACHINERY;

ITS PRINCIPLES AND MANAGEMENT

———

CHAPTER I

Introduction—Value of Refrigeration to Commerce—First Cargo of Frozen
Meat Imported — Magnitude of the Trade — Various Uses to which
Refrigerating Machinery is put—Development of the Import Trade
in Meat and Fruit — How Fruit should be packed — Principles of
Refrigerating Machinery—Laws of Heat—Rumford's Discovery—
English Unit of Heat—Joule's Equivalent—Latent Heat—Specific
Heat—Production of Cold—Absolute Zero.

THE use of refrigerating machinery has of late years
greatly increased in this country, on account of its having
become more and more dependent upon other countries,
or upon its own far-distant colonies, for its food supplies ;
and the establishment of the trade of transporting frozen
meat and other perishable food marks a new departure
in the history of modern commerce, which will probably,
in the near future, affect more or less all the over-
populated countries of Europe.

Twenty years ago such a trade would have been an
impossibility, as the mechanical devices for transporting
perishable products had not then been invented. It is
quite true that large quantities of tinned goods, including

I

beef and mutton from Australia, had been imported; but these were not favourably received. Inventors were therefore encouraged to devise some method by which the whole carcase of a slaughtered animal could be carried from the antipodes to this country, uncovered by tin or other metal, and simply surrounded by cold, dry air, reduced to a freezing point. This once achieved, the vast supplies of beef and mutton from Australia, New Zealand, and the River Plate would be available to feed the dense populations of our great cities. The margin of profit would be so great that the cost of transportation, and of working freezing machinery, would be readily borne by the promoters of such a trade, if once the necessary invention was procured.

It is quite true that refrigerating machines were made more than fifty years ago, but it was not until early in the year 1881, when the *Strathleven*, fitted with one of Bell - Coleman's machines, arrived here from Australia with the first cargo of frozen meat ever landed in this country. This cargo was followed by one brought over in the *Protos*, which was supplied with a cold-air machine made in Australia, and copied from one made on Lightfoot's patent, which had previously been sent to that colony.

In the autumn of the same year, the Orient Steam Navigation Company's steamship *Orient* also arrived in the Thames from Sydney, with a cargo of frozen beef and mutton in good condition, and by this time the problem of importing frozen meat may be said to have been practically solved. The refrigerating machinery on board the *Orient* was on the cold-air principle, and was the invention of Mr., now Sir A. Seale Haslam, of the Union

Foundry, Derby. This freezing machine was one of the first which had ever been put on board ship, and it worked, without even a momentary stoppage, during the six weeks' voyage home, the greater part of which was under a tropical sun. From that day to this the frozen provision trade has been steadily developing,—not in the meat department alone, for it now includes fish, milk, fruit, vegetables, etc.; and supplies of fresh perishable provisions are regularly received from Australia, New Zealand, the River Plate, Canada, the United States, and other parts.

The magnitude of the importation of fresh provisions into the United Kingdom is known and appreciated only by those persons who are brought into close relation with the trade, but the proportion that refrigerated meat bears to the whole quantity imported, and the enormous and rapid increase of that proportion, are known perhaps to but very few even of those engaged in it. The following statistics, in the compilation of which care has been exercised to obtain complete and reliable returns, will serve to illustrate the point.

The imports of fresh meat into the United Kingdom in the years 1886, 1887, and 1888 amounted to 260,000 tons, of which no less than 224,700 tons had been subjected to some method of artificial refrigeration; while in 1893 the total imports exceeded 218,500 tons, or nearly as much as the amount for the whole of the above three years. The rapid advance of this industry is shown from the fact that while in the year 1880 only 400 carcases of frozen mutton were imported, more than 1,993,000 were received in 1888, and this in 1893 had increased to over 2,500,000 carcases, of which nearly two-thirds were landed in London, one of the principal reasons for this port having the preference being that greater facilities for cold storage were to be found there. That other ports are, however, becoming alive to the advantages of the trade, and to the importance of providing proper cold storage accommodation, is evidenced by the recent establishment in Liverpool, Glasgow, and other ports of refrigerating stores ; but, no matter at what ports the cargoes may be discharged, the fact that inland conveyance is necessary to distribute the meat amongst retail purchasers requires no demonstration. This question of distribution to the various centres of the population has, however, so far been practically neglected, and the meat is still mainly carried in ordinary vehicles by road and rail, and Birmingham and Nottingham are the only inland towns in which cold stores on a large scale are at present available.

The condition in which the meat is received from the import vessels leaves little to be desired, and any deterioration in quality almost invariably occurs in transit from the import vessels to the cold stores on land, and between these stores and the retailer. During these times the meat is necessarily exposed to the temperature of the atmosphere, which is ordinarily many degrees higher than that at which the meat has to be kept to preserve it in good condition. Attempts have from time to time been made to overcome this difficulty, by providing specially constructed vans, artificially cooled by ice and salt ; but the great increase of weight entailed, the disadvantage occasioned by the moisture from the melting ice, the additional labour necessary, and the increased cost, have prevented the general adoption of such conveyances. In the United States, cars carrying vessels charged with a volatile liquid, which is allowed slowly to vaporise through pipes into other vessels, thereby reducing the temperature in the cars, have been used, but the danger of damage to the contents of the cars from leakage of the volatile liquid adds so greatly to the risk, that this system has not been adopted in this country.

The practical method of insuring the delivery of meat and other perishable provisions in good condition, is evidently to be found in vehicles so constructed that any desired temperature can be maintained for considerable periods at a small cost, without additional labour, the carrying of machinery, or a material increase in weight, and the contents of which are free from danger of damage by the ordinary risks of transit. That such means of transport can be provided has been demonstrated by the London and Tilbury Lighterage Company Limited, who have fitted up a fleet of barges that have been plying upon the Thames since early in 1888. Since that time they have conveyed many thousands of tons of frozen meat, all of which has been landed in good condition, and frequently at lower temperatures than when discharged from the import vessels. Such meat has realised the best prices when put upon the market, and it is now proposed to construct vehicles on the same principle for use by road and rail.

In addition to the vessels conveying frozen meat, there have now been established at all the ports of export and import large stores or cold-air chambers, in which the meat and other articles may be kept before being placed on board ships, or sent to the various markets for sale to the retailers. Refrigerating machines are now also used for cooling hospitals and public buildings in hot countries.

In breweries, for cooling the water to be used for refrigerating and attemperating, for cooling the air in the fermenting yeast rooms and store cellars, and, in addition, for the manufacture of ice for cooling the beer. In provision cooling stores, for the use of butchers, poulterers, fishmongers, and others. In dairies, for preserving milk, butter, and cheese, and for cooling milk in order to facilitate the formation of cream and butter, and also in the manufacture of artificial butter. In paraffin oil works, to enable refiners to extract in the presses a far larger quantity of paraffin than can be obtained in any other way,—at the same time, the quality of the oil separated is very much improved. In chemical works, for the reduction of mother liquors to low temperatures to increase the speed of crystallisation and the quantity of crystals produced, also for the freezing of various chemicals and other purposes. In bacon-curing works, for the production of mild - cured bacon, refrigerating machinery is also of the highest importance. In distilleries, for keeping the store tanks cool, and to prevent the very considerable loss which occurs by evaporation. In chocolate manufactories, for rapidly cooling the chocolate in the moulds. In dynamite factories, for the cooling of dynamite during the nitrating process. For numerous other processes, too numerous to detail, these machines are found to be invaluable, where a cold dry air is required, both in facilitating and quickening the process employed, and also in enabling them to be carried on at a reduced temperature. Special machines are also made for the establishments of wine growers, as well as for the manufacture of ice in tropical countries, in some of which it is now a large and important industry.

With a view to developing this new import trade by affording it the best facilities, the Peninsular and Oriental Steamship Company have had a number of their largest vessels fitted with refrigerating chambers and machinery. These chambers are so designed that they can be used either for frozen meat or fruit cargoes, as the market or season of the year may offer, thereby enabling them to utilise the machinery throughout the whole year. Of course, should neither of these perishable cargoes be obtainable, the space in the cold-air chambers can still be used for stowing ordinary merchandise. But the importance of having vessels equipped in such a way that the shipowner is able to accept any class of cargo, is doubtless the consideration which has impressed itself on the minds of the directors of the P. & O. Company, as it must eventually impress itself on the minds of shipowners generally. It may therefore be anticipated that refrigerating machinery and cool chambers will soon come to be regarded as part of the outfit of all large modern ocean-going vessels, whether sail or steam. The average carrying capacity of each of the P. & O. Company's ships thus fitted is 25,000 cases of fruit, or about 1300 tons. During the winter, meat can be carried home in a frozen condition from Australia and New Zealand, and from March to August the chambers will probably be filled with fruit.

The machinery in these P. & O. ships was originally designed for the frozen meat trade, extra large and conveniently arranged meat chambers being provided, and fruit chambers for special fruits. But, seeing the success with which fruit and vegetables could be preserved, the company decided to let the *Oceana* bring home a large quantity of fruit, chiefly Australian and Tasmanian apples. Five hundred tons of these were accordingly shipped from Melbourne on 5th May 1888, and when the vessel arrived at the Albert Docks the great bulk of them were in as prime a condition as when they were taken on board. This was the first large cargo of fruit brought from Australia, the beginning of what promises to be a new and important branch of commerce, and an encouragement for the Australian fruit-grower, such as he had not previously hoped for.

The apples were packed in boxes of 180 each, weighing about 50 lbs. The cold-air chamber, with a capacity of 20,000 cubic feet, in which they were stored, had been specially arranged by the engineers to adapt it for the conveyance of fruit as well as of meat. This was necessary, in view of the fact that, while meat can hardly be frozen too much while in course of transit through the tropics, the slightest touch of frost will spoil a consignment of fruit. In the case of the latter, the efforts of the engineer are directed to maintain a uniform temperature throughout the cargo, and, if possible, through every apple,—ranging from 45 to 55 deg. Fahr. In order to accomplish this, two air trunks are arranged along the sides of the storage chamber—one being fitted with openings through which the cold air enters, and the other situated at the opposite side, with orifices through which the same air is drawn away again by means of exhaust fans. Thus the air, reduced to the required temperature, is forced to circulate among the boxes of fruit, the object being to expose each box and its contents to the same degree of cold throughout the voyage. It follows that special care is necessary in packing the

fruit and in storing the boxes or cases in the best positions
in the cold-air chamber. The temperature on the day of
shipment was 65 deg. This had to be reduced throughout the cargo of fruit to 45 and 50 deg. To do this
the Haslam cold-air machinery was kept going only about
twelve hours per day in the warmer latitudes, and not
more than six hours daily after the Suez Canal was
passed. Throughout the thirty-nine days of the voyage,
the machine worked without a hitch.

Much has since been learnt from experience as regards
the stowage of the cases of fruit. They must not be
allowed to touch the sides of the chamber in such a way
that the circulation of the air is interfered with, nor can
the cases be placed directly one upon the other. Battens
or laths about $\frac{5}{8}$ of an inch in thickness are placed
between each tier of cases, and passage ways are left in
order that the condition of the fruit can be ascertained,
and any snow which may be created cleared away.
Much care is of course necessary in ensuring perfect
ventilation between the cases, and in maintaining the
requisite temperature. Some time might be saved in
stowage, if either the top or bottom of each case had two
half-inch battens nailed across it. This would do away
with the work of placing separate battens between every
two layers of cases, and would ensure the permanent
dunnage required for the access of cold air. The sizes
of the cases now used, about 3 feet by 18 inches by 10
inches, is quite convenient, the capacity averaging 40 lbs.
of apples, each of which is wrapped in a separate paper;
but the wood of the cases should be hard and tough. It
must be remembered that the cases are not always
handled very gingerly. Too little time is sometimes

allowed for stowing, for there is no doubt that the fruit would be less likely to deteriorate on the voyage if the cases were lowered into the hold with the minimum of bumps and bruises. It is certainly a mistake to shoot them like coal down a sharp incline, or to let them drop one upon the other.

It was in 1886 that West Indian fruit was first imported into this country, Messrs. Scrutton, Son, & Co., of London, having in that year had one of their ships, the *Nonpariel*, fitted with special chambers for storage, the Haslam dry-air system being the one adopted. This vessel on her first trip brought to London in splendid condition a large cargo of fruit, including West Indian bananas, in considerable quantities; sapsodillas, bel-apples, loquats, limes, pines, and strawberries from Madeira, all of which were purchased, almost immediately after arrival, by a Covent Garden merchant. On the voyage out, the *Nonpariel* called at Madeira, and there filled her freezing chamber with cabbages, French beans, peas, asparagus, carrots, and other vegetables, all of which were taken to Demerara, where they fetched good prices. So successful was the experiment that since then regular supplies of fresh fruit have been imported for the London market.

In addition to the above-mentioned companies, most of the large steamships trading to Australia and New Zealand are now fitted with refrigerating machinery, so as to carry cargoes of frozen meat or fruit when required; and nearly every passenger steamship afloat now carries a refrigerating machine for the purpose of preserving the fresh provisions carried for the use of passengers.

As the machines are placed under the charge of the engineers of the ship, each of whom is responsible for their continuous working while he remains on watch, it is obviously incumbent upon every seagoing engineer to make himself thoroughly acquainted with the principles and management of the different types of refrigerating machines. This work is intended to supply such information as will enable them to work machinery of this description in the most efficient manner, and the author will endeavour to state it in such plain terms that it may be easily understood by every engineer.

In order to thoroughly comprehend the principles of refrigerating machinery, it is necessary to acquire at least an elementary knowledge of the recognised definitions of heat, and of the laws which govern it.

Heat pervades all known substances in a greater or less degree, each substance having its own specific heat value. Heat can neither be created nor destroyed, but it may be transferred; for instance, when two bodies which have unequal temperatures are brought close together, there will take place between them a transfer of heat from the hotter of the two to the other, and the tendency towards an equalisation of heat or equilibrium of temperatures in this way is universal. The passage of heat takes place in three ways, viz. by radiation, by conduction, and by convection or carriage from one place to another by the heated currents.

All bodies possess the property of radiating heat, and in doing so the hotter body loses heat, and the colder body receives it by means of a process occurring in some intervening medium which does not itself become heated thereby. Heat rays proceed in straight lines, and the intensity of the heat radiated from any one source of heat diminishes as the distance from the source of heat increases, in the inverse ratio of the square of the distance. That is to say, for example, that at any given distance from the source of radiation, the intensity of the radiant heat is four times as great as it is at twice the distance, and nine times as great as it is at three times the distance.

Conduction is the flow of heat along or through an unequally heated body from places of higher to places of lower temperature, or from one substance to another in contact with it.

Convection is the motion of the hot body itself carrying its heat with it; for example, by the products of combustion in a furnace towards the heating surface in the flues of a boiler.

Up till the end of last century, heat was supposed to be a kind of matter which differed from all other forms of matter with which we are acquainted, in that it had no weight; but subsequent investigations proved this theory to be incorrect. In 1798, Benjamin Thompson, Count Rumford, an American who was then in the Bavarian service, presented a paper to the Royal Society of Great Britain, in which he stated the results of an experiment which he had recently made, proving the immateriality of heat and the transformation of mechanical into heat energy. His experiment consisted of the determination of the quantity of heat produced by the boring of a cannon at the arsenal at Munich. Rumford, after showing that this heat could not have been derived from any of the surrounding objects, or by compression of the materials employed or acted upon, says: "It appears to me extremely difficult, if not impossible, to form any distinct idea of anything capable of being excited and communicated in the manner that heat was excited and communicated in these experiments except it be motion." He estimated the heat produced to give the "mechanical equivalent" of the foot-pound as 783·8 heat units, differing but 1·5 per cent. from the now accepted value.

Had Rumford been able to eliminate all losses of heat by evaporation, radiation, and conduction, to which losses he refers, and to measure the power exerted with accuracy, the approximation would have been still closer. Rumford thus made the experimental discovery of the real nature of heat, proving it to be a form of energy, and, publishing the fact a half-century before the now standard determinations were made, gave us a very close approximation to the value of the heat equivalent. He also observed that the heat generated was exactly proportional to the force with which the two surfaces were pressed together, and to the rapidity of the motion.

The first absolutely conclusive experiment which established the fact that friction makes bodies hot, while it does not diminish their capacities for heat, was made by Sir Humphry Davy shortly afterwards (1799), and conclusively confirmed these deductions from Rumford's work. His experiment consisted in rubbing together two pieces of ice till they melted into water, due care having been taken to prevent heat from entering the ice by any other means than by friction alone. Davy thereupon concluded : "It is evident that ice by friction is converted into water. . . . Friction, consequently, does not diminish the capacity of bodies for heat." He then, in 1812, for the first time, stated plainly and precisely the real nature of heat, saying : " The immediate cause of the phenomenon of heat, then, is motion, and the laws of its communication are precisely the same as the laws of the communication of motion."

Heat and mechanical force therefore are identical and convertible.

Independently of the medium through which heat may be developed into mechanical action, the same quantity of heat is resolved into the same total quantity of work.

The English unit of heat is that which is required to raise the temperature of 1 lb. of water 1 deg. Fahr. If 2 lbs. of water be raised 1 degree, or 1 lb. be raised 2 degrees in temperature, the expenditure of heat is the same in amount, viz. 2 units of heat; and to express the mechanical equivalent of heat, the comparison lies between the unit of heat on the one part, and the unit of work or the foot-pound on the other part. The most precise determination yet made of the numerical relation subsisting between heat and mechanical work was obtained by Dr. Joule, who found through turning a paddle wheel, by a definite power in a known weight of water, that the heat communicated amounted to 1 deg. Fahr. for every 772 foot lbs. of work expended in producing it. The mechanical equivalent of heat known as Joule's equivalent is therefore taken as 772 foot lbs. for 1 unit of heat.

It has been proved repeatedly that the expansion of air or other gas into a space, without doing work, produces no fall in temperature, but that, on the contrary, if the air at a temperature say of 230 deg. Fahr. be expanded against an opposing pressure or resistance, as against the piston of a cylinder, giving motion to it or otherwise doing work, the temperature will fall nearly 170 deg. Fahr. when the volume is doubled, *i.e.* from

230 deg. Fahr. to about 60 deg. Fahr., and taking the initial pressure at 40 lbs., the final pressure would be 15 lbs. per square inch.

The heat which is required to melt a certain quantity of a solid at the melting point into a liquid, at the same temperature, is called the latent heat of fusion.

It is called *latent* heat, because the application of this heat to the body does not raise its temperature or warm the body.

For instance, if we apply heat to a pound of ice at 32 deg. Fahr., it will begin to melt, but the temperature will remain stationary till the whole of the ice is turned into water, and to effect this transformation, 142 units of heat must be supplied. As the temperature remains stationary during the melting process, the question arises, What becomes of the heat which has been expended? The early discoverers of this phenomenon being unable to account for the heat thus apparently lost, invented the theory that it had become *latent*, or concealed in the water, and, in accordance with this theory, it was said that the *latent heat* of water was 142 deg. In accordance with the mechanical theory, it is recognised that the heat thus expended is spent in doing internal work on the particles of the ice, which results in their cohesion being overcome, so that the condition of the ice is changed from the solid to the liquid state. We should say, therefore, that 142 units of internal work, or of latent work, are done upon the ice in order to transform it into water.

The specific heat of a body signifies its capacity for heat, or the quantity of heat required to raise the temperature of a quantity of water of equal weight 1 deg.

The British unit of heat is that which is required to raise the temperature of 1 lb. of water 1 deg. from 32 to 33 deg. Fahr., and the specific heat of any other body is expressed by the quantity of heat in units necessary to raise the temperature of 1 lb. weight of such body 1 deg.

The specific heat of water is represented by 1 or unity, and there are very few bodies of which the specific heat equals or exceeds that of water.

Liquids in the course of being transformed into vapour on the application of heat absorb a certain quantity of heat which remains latent in the vapour, and is, on the contrary, restored to sensibility and communicated to other bodies when the vapours are condensed into liquids.

As already mentioned, heat pervades all known substances, and it cannot be either created or destroyed; consequently, when we say a body is cold, we mean to say that it is cold in comparison with some other body; for example, if we take three basins containing water at the temperatures of 32 deg., 80 deg., and 130 deg. respectively, and place the right hand say in the water at 130 deg. and the left hand in the water at 32 deg., keeping them there for some little time; then placing *both* hands in the water at 80 deg., the right hand will feel the water cold and the left hand warm, thus proving that cold is only a relative term, or, in other words, that a body is rendered cold by the withdrawal of heat.

The production of cold, or, in other words, the abstraction of heat, is a curious subject of inquiry. When a salt is dissolved in water cold is produced. When a liquid vaporises, the heat, sensible and latent, necessary for the production of the vapour is abstracted from some other body in contact with the liquid, and cold is produced. When spirits of wine or aromatic vinegar, for example, is thrown on the body, a sense of cold immediately results from the vaporisation of the liquid, which draws heat from the body. If air is allowed to expand, there is a reduction of temperature, and a translation of heat from neighbouring bodies. Again, in hot climates, water is successfully cooled in porous vessels, through the pores of which the water passes to the exterior, and is vaporised, and the cooling process is accelerated by a current of air directed upon the vessel which quickens the vaporisation. A similar process for cooling is carried out by the medium of moss, grass, and other like substances.

For the production of intense cold, mixtures of various salts and acids in varying proportions with water are very effective, but more intense degrees of cold are produced with snow or ice.

The most intense cold as yet known was produced by Professor Faraday during his experiments on the liquefaction and solidification of gases, by the evaporation of a mixture of solid carbonic acid and sulphuric ether under the receiver of an air-pump. He found that when a perfect vacuum was nearly approached, an intense cold, measured by —166 deg. Fahr. or 166 deg. below zero on F scale, was obtained.

Absolute zero—the point at which heat motion ceases —or, in other words, when all the heat is abstracted from a body, has been found by experiment to be 493 deg. Fahr. below the freezing point of water, or 461 deg. below zero on the F scale.

2

At the risk of repetition, it may be advisable here to briefly recapitulate and emphasise the more important points already explained. As heat pervades all substances in a greater or less degree, and can neither be created nor destroyed, and as a body is rendered cold simply by the transfer of a portion of its heat to some other substance, it follows that all refrigerating machines are but devices for effecting the transference of heat, and are consequently governed by its laws. This is the reason why we have devoted so much space to the leading definitions and laws of heat, and no doubt the reader will fully appreciate the importance of acquiring a proper understanding of that subject.

CHAPTER II

ALTHOUGH it is only within recent years that the theory
of Refrigeration has been properly understood, artificial ice
has, by more or less primitive methods, been produced
for centuries.   The ancient Egyptians employed for this
purpose earthenware vessels filled with water, which was
evaporated by fanning it with a palm-leaf.   Even at the
present time, ice is manufactured in India by digging
troughs in the ground about a foot deep, which are lined
with rice-straw or some other non-conducting substance,
and upon this placing shallow porous earthenware vessels
filled with water.

The manner in which the ice is formed may be explained as follows : When a liquid changes its state into
that of a gas, the transformation is effected by the
addition of energy in the form of heat, and this effect

19

may be produced without change in the sensible temperature, provided the heat is absorbed at the same rate as it is supplied from without.

The air in India being extremely dry, the tendency of the water to evaporate is naturally very strong. In consequence of the rapid evaporation from this cause, heat is absorbed at a greater rate than it can be supplied from the surrounding atmosphere, the heat in the water is therefore drawn upon and its temperature consequently falls to below freezing point, with the result that a thin layer of ice is soon formed. This layer is at once removed, and the process repeated as often as required.

We will now proceed to describe the various systems of refrigerating machinery that have from time to time been introduced, showing their development, their practical application, and methods of working.

The primary function of all refrigerating and ice-making apparatus is to abstract heat, the temperature of the refrigerating agent being of necessity below that of the substance to be cooled. It is obvious, however, that without provision either for rejection of the heat thus abstracted, or for renewal of the refrigerating agent, equalisation of temperature would ultimately ensue and the cooling action would cease. In practice, if the machine is to work continuously, one or other of these means must be adopted; and a complete refrigerating machine therefore consists of an apparatus by which heat is abstracted, in combination either with some system for renewing the heat-absorbing agent, or, as is more usually the case, with a contrivance whereby the abstracted heat is rejected, and the agent restored to a condition in which it can again be employed for cooling purposes.

When a substance changes its physical state, and passes from the solid to the liquid form, the force of cohesion is overcome by the addition of energy in the form of heat. The effect may be produced without change in sensible temperature, if the heat be absorbed at the same rate as it is supplied from without.

Thus, as is well known, the temperature of melting ice remains constant at 32 deg. Fahr.; and any increase or decrease in the heat supplied merely hastens or retards the rate of melting, without affecting the temperature. Mixtures of certain salts with water or acids, and of some salts with ice, which form liquids whose freezing points are below the original temperatures of the mixtures, do not however behave in this way; for, under ordinary circumstances, the tendency to pass into the liquid form is so strong that heat is absorbed at a greater rate than it can be supplied from without. The store of heat of the melting substances themselves is therefore drawn upon, and the temperature consequently falls until a balance is set up between the rate of melting and the rate at which heat is supplied from outside. This is what takes place with ordinary freezing mixtures. The amount of the depression in temperature appears to depend to some extent on the state of hydration of the salt, and the percentage of it in the mixture. Almost the only salts used are those of certain alkalies, few others possessing the requisite solubility at low temperatures. A list of freezing mixtures usually employed is given in the Appendix, Table A.

Such a method of abstracting heat is extremely convenient for the laboratory, and for some other special purposes. Attempts have also been made to apply it commercially on a large scale for the manufacture of ice and for cooling. The late Sir William Siemens constructed an ice-making apparatus in which calcium chloride, commonly known as chloride of lime, was employed. The reduction in temperature produced by dissolving this salt in water is about 30 deg. Fahr.; but as this was not sufficient for freezing when the initial temperature of the water was about 60 or 65 deg. Fahr., a heat interchanger was introduced, by means of which the spent liquor at about 30 deg. was utilised for cooling the water before it was mixed with the salt; and to the extent of this cooling the degree of cold produced was intensified. The salt was recovered by evaporation, and used over again. Although this apparatus worked well and produced ice, the inventor himself considered the process inferior to mechanical methods and abandoned it.

In the Toselli machine a mixture of ammonium nitrate and water is used, by means of which a reduction in temperature of about 140 deg. Fahr. is obtained. The apparatus consists of a vessel in which the solution of the salt is effected, and an ice can containing several slightly tapering moulds of circular cross section and of varying sizes. The moulds being filled with water are introduced into the freezing mixture; and in a few minutes ice is formed round the edges to the thickness of nearly an eighth of an inch. The rings or tubes of ice are then removed and placed one within the other, and so form a small stick of ice.

Ammonium nitrate is also employed in a machine recently brought out in the United States for the production of ice on a large scale. In one form of this apparatus, intended chiefly for domestic purposes, a series of annular vessels, one within the other, is used, the moulds in which the ice is to be formed being placed in the centre. The reduction of temperature produced by the freezing mixture in the outermost vessel cools the water in the second, and this, on salt being added, cools the third, and so on. In this way the cold is very much intensified at the centre, and a low temperature can be produced independent of the initial temperature of the water. The number of rings employed varies according to the effect to be produced and the conditions under which the apparatus is applied. The annular vessels, together with the ice moulds, are carried in a wood casing supported on bearings, the only motive power required being that necessary to rotate the vessels slowly, so as to promote the solution of the salt.

Another form of apparatus, designed for continuous use on a large scale, consists of a vessel into which ammonium nitrate is automatically fed, and in which it enters into solution with water previously cooled in an interchanger by the spent liquor, after the latter has left the ice-making tanks or cooling rooms. The cold brine thus produced is circulated by a pump through the ice tanks, or through pipes placed in the rooms it is desired to cool; and is returned through the interchanger to an evaporating tank, where by means of heat the water is driven off and the salt recovered. This is practically Sir William Siemen's apparatus in a somewhat extended form. The cost of producing 15 tons of ice per twenty-four hours with such an apparatus of suitable capacity is stated at 7s. per ton, estimating the coals used at 15s. per ton, but excluding, of course, depreciation and repairs of machinery. This, however, is rather too low an estimate, being based upon the assumption that 1 lb. of coal is capable of evaporating 20 lbs. of water. Nearly the whole of the coal is expended in evaporating the water in recovering the salt, the quantity being given at 2½ tons of coal for every 15 tons of ice. This, however, being calculated upon an evaporative duty of 20 lbs. of water per pound of coal, the amount actually used would probably be about 5 tons of coal, which would make the cost per ton of ice 9s. 3d. instead of 7s. On the other hand, it must be remembered that, under certain climatic conditions, much of the water could be evaporated in the open air, without the use of fuel ; in which case the coal consumption, and therefore the cost of ice production, would be considerably lessened.

In all cases where a liquid is employed, the refrigerating action is produced by the change in physical state from the liquid to the vaporous form. It is, of course, well known that such a change can only be brought about by the acquirement of heat, which is absorbed in increasing the energy of the molecules. This heat, which is absorbed without changing the sensible temperature, is usually called the latent heat of vaporisation. For the purpose of refrigeration, by which must be understood the abstraction of heat at temperatures below the normal, it is obvious that, other things being equal, that liquid is the best which has the highest heat of vaporisation, because with it the least quantity has to be dealt with in order to produce a given result, and therefore the power expended in working the machine will be the least. In fact, however, liquids vary, not only in the amount of heat required to vaporise them, but also according to the temperature or pressure at which vaporisation occurs, and in addition to this they vary in the conditions under which such a change can be effected

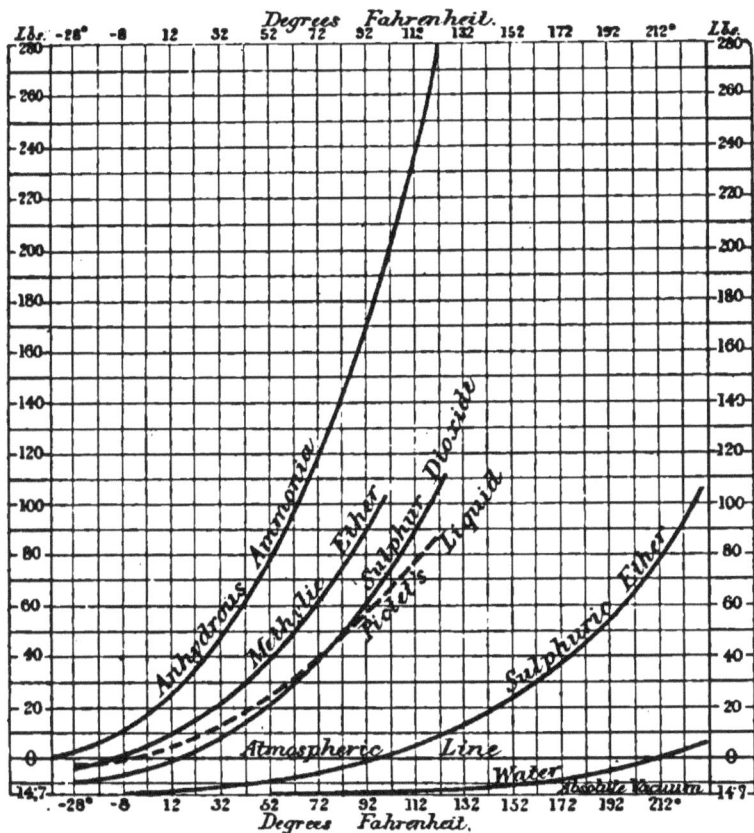

Fig. 1.

For instance, water has an extremely high latent heat, but as its boiling point at atmospheric pressure is also high, evaporation at such temperatures as would enable it to be used for refrigerating purposes can only be effected under an almost perfect vacuum. This is graphically shown in Fig. 1. The boiling point of anhydrous ammonia, on the other hand, is 37·5 deg. below zero Fahr. at atmospheric pressure, and therefore, for all ordinary cooling purposes, its evaporation can take place at pressures considerably above that of our atmosphere. The curve of vapour tensions of anhydrous ammonia is also given in Fig. 1, as well as those of some other agents which are used for refrigerating purposes, namely, methylic ether, Pictet's liquid, sulphur dioxide, and ether. In connection with this it may be mentioned that Pictet's liquid is a compound of carbon dioxide and sulphur dioxide, and is said to possess the property of having vapour tensions, not only much below those of pure carbon dioxide at equal temperatures, but even below those of pure sulphur dioxide at temperatures above 78 deg. Fahr.

The considerations, therefore, which chiefly influence the selection of a liquid refrigerating agent are—

1. The amount of heat required to effect the change from the liquid to the vaporous state, commonly called the latent heat of vaporisation.

2. The temperatures and pressures at which such change can be effected.

This latter attribute is of twofold importance ; for in order to avoid the renewal of the agent, it is necessary to deprive it of the heat acquired during vaporisation, under such conditions as will cause it to assume the liquid form, and thus become again available for refrigeration. As this rejection of heat can only take place if the temperature of the vapour is somewhat above that of the cooling body which receives the heat, and which for obvious reasons is in all cases water, the liquefying pressure at the temperature of the cooling water, and the facility with which this pressure can be reached and maintained, is of great importance in the practical working of any refrigerating apparatus.

Now, seeing that the liquid is produced by cooling by means of water, which may have an initial temperature as high as, say, 90 deg. Fahr., it is obvious that a part of the cold produced in evaporation must be expended in reducing the temperature of the liquid itself from the temperature at which it leaves the condenser down to the temperature of the refrigerator, which may be as low as, say, zero Fahr., or even lower. Consequently, it is of vital importance that the refrigerating agent should give off relatively little heat in being cooled, or, in other words, that its specific heat should be low in proportion to its latent heat of vaporisation. Suppose, for instance, that we have a liquid with a specific heat of 1, that is to say that for every degree Fahr. through which the liquid is cooled, each pound gives off one thermal unit, and that its heat of vaporisation is 150 thermal units—then if the refrigerator temperature is zero Fahr., and the liquid leaves the condenser at, say, 100 degrees, it is obvious that 100 units of heat will have to be extracted from every pound of liquid to reduce its temperature to that of the refrigerator; and, as the total heat required in evaporating the one pound of the liquid is 150 units, it follows that the available useful refrigerating work per pound of liquid will be 150 minus 100, or only 50 units. In other words, 66 per cent. of the whole refrigerating power is uselessly expended in cooling the liquid itself.

This example will suffice to show that it does not necessarily follow that a liquid with a high latent heat of vaporisation is a good refrigerating agent, or that one with a low heat of vaporisation is bad. What is required is that the difference between the latent heat of vaporisation and the temperature at which the liquid leaves the condenser, minus the refrigerator temperature multiplied by the specific heat of the liquid, is large.

Other matters of importance in the selection of a liquid refrigerating agent are, the pressures at which vaporisation and condensation occur at those temperatures met with in practice, besides which there are the questions of inflammability and stability of the chemical compound. This latter is of considerable importance, more especially in relation to the successful working of machinery in hot climates, because more than one refrigerating agent now before the public, though fairly suitable for use in temperate climates, is quite unsuited for use in the tropics, owing to its liability to alteration at high temperature. It does not follow, therefore, that a machine which works fairly well under one set of conditions, will answer its purpose when the conditions are changed.

The principal systems in which the evaporation of liquids is employed may be treated under the following subdivisions :—-

A. Apparatus in which the refrigerating agent is rejected along with the heat it has acquired.

B. Machines in which heat only is rejected, the refrigerating agent being restored to its original physical condition by means of mechanical compression, and by cooling when under compression.

C. Apparatus in which heat only is rejected, by allowing the refrigerating agent to change its physical condition by entering into solution with a liquid, from which it is afterwards separated by evaporation, and recovered.

D. Machinery and apparatus in which heat only is rejected, by changing the physical state of the refrigerating agent by means of a combination of both mechanical compression and solution, with cooling.

System A. This is generally known as the vacuum process; for, as the refrigerating agent itself is rejected, the only agent of a sufficiently inexpensive character to be employed is water, and this, owing to its high boiling point, requires the maintenance of a high degree of vacuum in order to produce ebullition at the proper temperature.

In Fig. 1 are shown graphically the vapour tensions of water at temperatures up to boiling point at atmospheric pressure, the actual figures being given in the Appendix, Table B; from which it may be observed that at 32 deg. Fahr., the tension is only 0·089 lbs. per square inch. In ice - making, therefore, a degree of vacuum must be maintained at least as high as this.

The earliest machine of this kind appears to have been made in 1755 by Dr. Cullen, who produced the vacuum by means of an air pump. In 1810, Leslie, combining with the air pump a vessel containing strong sulphuric acid, for absorbing the vapour from the air drawn over, and so assisting the pump, succeeded in producing an apparatus by means of which 1 to 1½ lbs. of ice could be made in a single operation. Vallance and Kingsford followed later, but without practical results; and Carré, many years afterwards, embodied the same principle in a machine for cooling and for making small quantities of ice, chiefly for domestic purposes. His machine, which is still sometimes used, consists of a small vertical vacuum pump worked by hand, either by a lever or by a crank, which exhausts the air from the carafe or decanter containing the water or liquid to be frozen or cooled. Between the pump and water vessel is a lead cylinder, three-quarters full of sulphuric acid, over which the air, and with it the vapour given off from the liquid, is caused to pass on its way to the pump. The vacuum thus produced causes a rapid evaporation, which quickly lowers the temperature of the water; and, if the action is prolonged for about four or five minutes, the water becomes frozen into a block of porous opaque ice. The charge of acid is about 4½ pints, and it is said that from fifty to sixty carafes of about a pint each can be frozen with one charge. So long as the joints are all tight, and the pump is in good order, this apparatus works well; but in practice it has been found troublesome and unreliable, and consequently has never come into anything like general use.

In 1878, Franz Windhausen, of Berlin, patented a compound vacuum pump for producing ice direct from water on a large scale, without the employment of sulphuric acid; and also an arrangement in which sulphuric acid could be used, the acid being cooled by water during its absorption of the vapour, and afterwards concentrated, so that a fresh supply was rendered unnecessary. This apparatus was improved on in 1880; and in 1881 a machine nominally capable of producing from 12 to 15 tons of ice per twenty-four hours was put to work experimentally at the Aylesbury Dairy, at Bayswater, being afterwards removed to Lille Road. The installation consists of six slightly tapered ice-forming vessels of cast-iron, of circular cross section, closed at their bottom ends by hinged doors, with air-tight joints, into which water is allowed to flow at a regular rate through suitable nozzles, the cylinders being steam-jacketted in order to allow the ice to be readily discharged. The upper parts of these vessels communicate with the pump, through a long horizontal iron vessel of circular section containing sulphuric acid, which, when the machine is in operation, is kept in continual agitation by means of revolving arms. The acid vessel is surrounded with cold water, which carries off most of the heat liberated during the absorption of the vapour. The pump has two cylinders, one double-acting, of large size, and a smaller single-acting one. The capacities of these cylinders per revolution are as 62 to 1. The air and whatever vapour has passed the acid are drawn into the large pump, which partially compresses and delivers them into a condenser. Here part of the vapour is condensed by the action of cold water, the remainder passing along with the air to the second pump, where they are compressed up to the atmospheric tension and discharged. The advantage gained by the use of a compound pump is due to the action of the intermediate condenser, and to the compression being performed in two stages, by which the losses from the clearance spaces in the large pump are rendered much less than they would be if compression to atmospheric pressure were accomplished in a single operation. The effect of the pump is said to be such that a vacuum of half a millimetre of mercury, or about 0·0097 lb., per square inch can be continuously maintained; though in actual work about $2\frac{1}{2}$ millimetres, or 0·0484 lb., per square inch is as low as is necessary. The concentration of the acid is effected in a lead-lined vessel, in which is a coil of lead piping heated by steam, the pressure in the vessel being kept down by means of an ordinary air-pump. No acid pump is needed, as the transfer from one vessel to another is effected by the pressure of the atmosphere. The comparatively cool and weak acid on its way to the concentrator is heated in an interchanger by the strong acid returning from the concentrator. Six blocks of ice, each weighing about 560 lbs., are formed in about sixty minutes after starting. The charge of acid is said to serve for four makings of ice, after which it becomes too weak, and requires to be concentrated.

The water being admitted into the ice-forming vessels in fine streams offers a large surface for evaporation, and is almost immediately converted into small globules of ice, which fall to the bottom and become cemented together by the freezing of a certain quantity of water that collects there. This water being in a violent state of ebullition, the ice so formed is not solid, but contains spaces or blow holes, which, as soon as the block is discharged from the vessel, become filled with air and cause opacity. Several attempts have been made to produce transparent ice by the direct vacuum process, but so far without success. Distilled water or water deprived of air has been tried, and hydraulic pressure has been used for compressing the porous opaque blocks ; but neither plan has been found practicable commercially. It would appear, indeed, that the only way to make clear ice by the vacuum process is by forming it in moulds or cells, subjected externally to the action of brine previously cooled by the evaporation of a portion of its water. The cost in this case would necessarily be greater; but the ice would be solid and transparent, and would consequently have a higher - commercial value.

The latent heat of liquefaction of water being 142·6 deg., at 32 deg. Fahr. the total heat to be abstracted in order to produce 1 ton of ice from 1 ton of water at 60 deg. Fahr., or 28 deg. above 32 deg. is $(142\cdot6° + 28°) \times 2240$, or 382,144 units. Taking the latent heat of vaporisation of water at 32 deg. Fahr. to be 1091·7, it is obvious that 350 lbs. $= \frac{382144}{1091\cdot7}$ must be evaporated to make the ton of ice. But, in addition, the sensible heat of the evaporated water, which, entering at 60 deg. would leave at about 32 deg., would have to be taken off; and this would require the evaporation of about $9\frac{1}{4}$ lbs. more, making a total of about 360 lbs., without allowance for loss by heat entering from without, which would be considerable. The total water actually used is given by Mr. Pieper at from 10 to 12 tons per ton of ice, including the quantity required for cooling purposes. The fuel consumption is stated to be 180 lbs. of coal per ton of ice ; but it is understood that a much larger quantity of coal is actually required, the excess being consumed in generating steam for driving the vacuum

3

pump and the concentrator air pump, and for evaporating the water absorbed by the acid.

According to Dr. Hopkinson, the cost of making 1 ton of opaque ice is 4s.; but experience has shown that a much higher figure is required to cover the necessary expense for repairs and maintenance, which in some parts of the apparatus are very heavy.

Windhausen's machine has not met with any extended application in this country, owing no doubt to the opaque and porous condition of the ice produced by it, and to the large and cumbrous nature of the plant, which must doubtless require great care and supervision in working.

In 1878, James Harrison patented a vacuum apparatus for refrigerating liquids by their own partial evaporation, and for making ice. Its chief feature was the revolving cylinder or pump (shown in section in Fig. 2), which affords a simple and efficient means of exhausting large volumes of vapour of low tension, without incurring the loss from friction of ordinary piston packings,

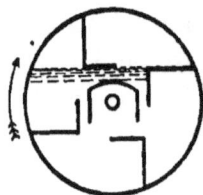

Fig. 2.

and the trouble of keeping them tight and in good working order, while at the same time the first cost is much reduced. The pump consists of a hollow iron cylinder, revolving on a horizontal axis, and divided into compartments by longitudinal partitions of L section. It is partially filled with a non-evaporable liquid, or one which evaporates only at a temperature considerably in excess of that at which the refrigerating liquid is evaporated, and which is also chemically neutral to the vapour that is brought in contact with it. In practice oil is the liquid used. The refrigerating or ice-making vessels, of any convenient form, are connected by a pipe with one end of a fixed hollow axle on which the cylinder revolves; and inside the cylinder another pipe rises up above the level of the liquid, the longitudinal partitions being stopped short at one end to enable this to be done. The compartments move round, mouth downwards, carrying with them the vapour with which they are charged, and compressing it to an extent measured by the distance they dip below the surface of the liquid; until, when the lowest position is approached, the compressed vapour is liberated, and rises into a fixed hood near the centre, in communication with a second hollow axle at the opposite end of the cylinder to that at which the vapour enters. Through this second axle the compressed vapour passes to a surface evaporative condenser, in which it is partly condensed by the combined action of direct cooling and of the partial evaporation of water trickling over the surface : the water of condensation, together with any air, being then compressed to the tension of the atmosphere by a small pump, and then discharged. By this process, the author is informed, it is expected to produce opaque ice on a large scale, at a cost of about 1s. per ton. The fuel consumption will certainly be very small, because friction, which is a large item in the Windhausen apparatus, is here to a large extent eliminated. There would also be a saving of all the fuel used in concentrating the acid, and of much of the water required for cooling purposes, besides a reduction in the first cost of the plant, and in the expense of maintenance.

Messrs. Southby & Blyth have also patented a vaccum

ice machine of a new and simple design.   In Fig. 3, A is

Fig. 3.

the cylinder of the large or vapour pump, single acting;

Aa the steam jacket round vapour pump which prevents the condensation of the vapour in the vapour pump ; B the chamber containing the crank and counterpoised fly-wheel which works the vapour pump A ; Bb the cover of crank chamber, hung on hinges at bottom ; C fast and loose driving pulleys ; D exit valve chest and condenser ; E starting gear handle ; F ejector pump, double acting, actuated by the eccentric ; G vacuum gauge ; H H valves to open or shut off either the ice box or cooling vessel as required ; K ice box containing the ice pail Kk surrounded by an air chamber ; L the automatic water feeder to ice pail, actuated by the eccentric.

The single-action vapour pump piston in the cylinder A, worked by the crank and flywheel contained in the chamber B, and driven by the pulley and shaft C passing through a stuffing box, draws the vapour from the water in the ice box or cooling vessel, through a valve in the piston, and delivers the compressed vapour, which is pre-vented from condensing in the cylinder by the heat of the steam jacket Aa, through the exit valve in the chest D, into the condenser, where it is condensed by a current of cold water, and from which it is drawn as water by the air pump F, together with any air that may have leaked into the machine.

The drawing off of the vapour from the ice box or cooling vessel causes the water to evaporate rapidly until it has given off all its latent heat, when it turns to ice ; but, before the heat of the water is drawn off, all the air must be exhausted out of both the machine and the water, and to do this with the large pump would require enormously more power than to drive off and compress the vapour which arises from the water in the ice box ;

but, by screwing in the starting screw E, the inlet valve in
the piston is kept open for any desired portion of the first
part of the delivery stroke of the pump A, allowing the
air thereby to return to the underside of the piston, thus
reducing and regulating, as may be desired, the power
required in getting up the vacuum. The action of the
machine is very simple, the little air pump F pumps away
the air out of the whole interior of the machine. This
vacuum fills itself with water vapour from any vessel
containing water put in connection with the machine.
The large pump A then compresses the vapour contained
in it; this compression of the vapour increases its temperature, and would cause it to condense into water in
the cylinder if it was allowed to lose its heat to the
cylinder, and in this case the condensed vapour to be
ejected would be so small in quantity that it would not
be forced through the exit valve. To prevent this loss
of heat of the compressed vapour and its consequent condensation into water in the cylinder, Messrs. Southby &
Blyth heat the cylinder to a temperature above that at
which the vapour will condense, and in this way the
compressed vapour can be almost entirely forced through
the exit valve into the condenser D. There it is condensed into water, and from it is drawn and ejected into
the air by the air pump, together with any air that may
have found ingress into the machine. It will be seen
from this description that the two points of this invention which cause its success, and without which little or
no ice would be made, are—(1) The use of two pumps, one
a small one to relieve the whole of the machine of the
pressure of the atmosphere, the other a very much larger
one to compress the vapour arising from water contained

in a vessel in connection with the machine, and so to force it into a condenser; (2) the heating of this larger cylinder, so that the compressed vapour may not condense in the cylinder.

As to the power required for driving with so rarefied a fluid as the vapour of water at low temperatures, the pressure on the piston is of course very small. The initial pressure of the vapour as it rises from the water at a freezing temperature being 0·185 inch, or, say, 0·15 inch in the cylinder, and the final pressure that will cause it to condense on a condenser cooled by water passing away at 100 deg. would be 2 inches, which would give an average pressure of about one-sixth of a pound on the piston. With so small a maximum pressure very light piston rings are all that are necessary, so that with a well lubricated piston the friction is very small indeed, and tests of the power used show this, we are informed, to be very small, and remarkably so for small machines, and there is, of course, no other expense, as it is free from all expense for chemicals.

Some minor details conduce to the success of the machine, more especially as regards starting it. When starting the machine, air at a comparatively high pressure has to be dealt with, occasioning an adverse pressure on the piston of, say, 7½ lbs., or some forty-five times that of the working pressure. Then the air, being non-condensible, will not disappear on compression in the condenser, as is almost the case with water vapour. The difficulty has, however, been overcome by using a bye-pass valve, opening a communication between the two sides of the piston, and capable of being closed at any point of the stroke, so that when starting this valve is kept open all but, say, for $\frac{1}{20}$th at the end of the stroke, and is closed at earlier points of the stroke as a more perfect vacuum is obtained. It is kept closed altogether when all the air has been expelled. All the joints of the valves, as well as of the covers, are of peculiar construction, and consist of a narrow shallow bead round the valve, or opening for the valve, closing on a more or less elastic bed, such as vulcanised india-rubber or fibre, which has been found to answer well. The valves also have to be of very large capacity to enable the machine to work at high speeds. The water, after being cooled, may, as in the case of cold store chambers, be in pipes, etc., any distance from the machine, and salt water may be used in some cases with economy and advantage.

The makers specially intend the machines for ice-making on board ship, or in confined places, or where the escape of injurious gases would be dangerous, and for making ice by hand power. The quantity of cooling water for the condenser is very small indeed, and it is stated that, by the addition of an extra cylinder and some increase to the power required for working, the cooling water can be dispensed with altogether. It is also noteworthy that the production of ice commences within a few minutes after starting the machine. A machine placed where most convenient, in a brewery, for instance, can cool vats of water for refrigerating or attemperating in various parts of the brewery. For working cold chambers, ice may be produced in sealed vessels, placed along the top of the cold chamber, by a machine worked at a distance, and only connected to the sealed vessels by a pipe, in which case the ice machine need be worked only when the ice in the vessel is nearly dissolved and requires to be refrozen.

System B. This is known as the compression process, and is used with liquids whose vapours condense under pressure at ordinary temperatures. Although prior to 1834 several suggestions had been made with regard to the production of ice and the cooling of liquids by the evaporation of a more volatile liquid than water, the

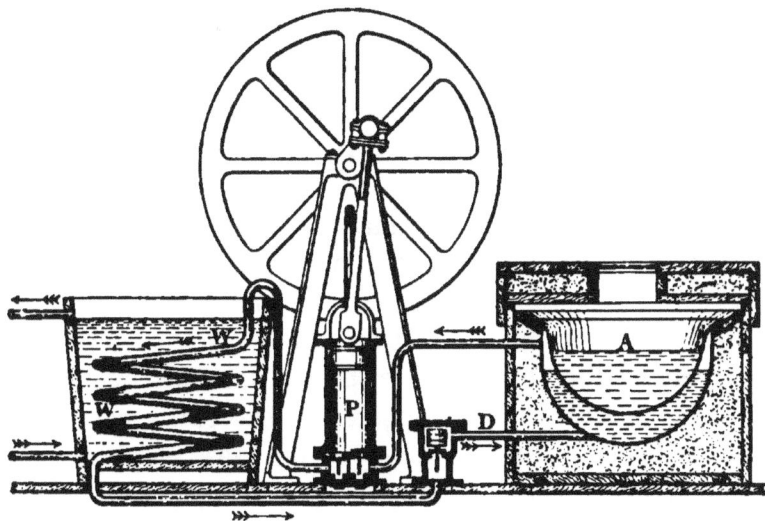

Fig. 4.

author believes that the first machine really constructed and put to work was made by John Hague in that year, from the design of Jacob Perkins.

According to Sir Frederick Bramwell, the liquid used was one arising from the destructive distillation of caoutchouc. The machine, which is shown in section in Fig. 4, so far as the author is aware, was never put to work outside of the factory where it was constructed.

The water to be frozen was placed in a jacketed copper pan A, the jacket being partially filled with the volatile liquid, and carefully protected on the outside with a covering of non-conducting material. A pump P drew off the vapour from the jacket, and delivered it compressed into a worm W, around which cooling water was circulated, the pressure being such as to cause liquefaction. The liquid collected at the bottom of the worm, and returned to the jacket through a pipe D, to be again evaporated. The apparatus, though in some respects crude, is yet the parent of all compression machines used at the present time, the only improvements made since the year 1834 having been in matters of constructive detail.

The next advance was made in 1856 and 1857 by James Harrison, who brought out a machine embodying the same principles as that of Perkins, but worked out on a larger and more practical scale. This machine was constructed for the inventor by the late Mr. Siebe, and was the first ice-making apparatus that really came into practical use in this country, and was employed for commercial purposes. An improved apparatus of this kind, in which sulphuric ether is used as the refrigerating agent, is still manufactured by Messrs. Siebe, Gorman, & Co.

Pure ether at ordinary temperature and atmospheric pressure, is a colourless, transparent, and very volatile liquid ; its taste is at first fiery, but afterwards cooling, owing to its rapid vaporisation, and it has an agreeable odour. It is lighter than water, its density in a liquid state being 0·723 at a temperature of 60 deg. Fahr. One peculiarity of ether is that, although one of the lightest of liquids, its vapour density is very great, being 2·586 times as heavy as air. Its boiling point at atmospheric pressure is 96 deg. Fahr., having a latent heat of vaporisation of 165, and it is solidified at a temperature of 24 deg. below zero Fahr., or 56 deg. below freezing point.

There are several methods by which ether may be prepared, but it is usually obtained by the distillation of

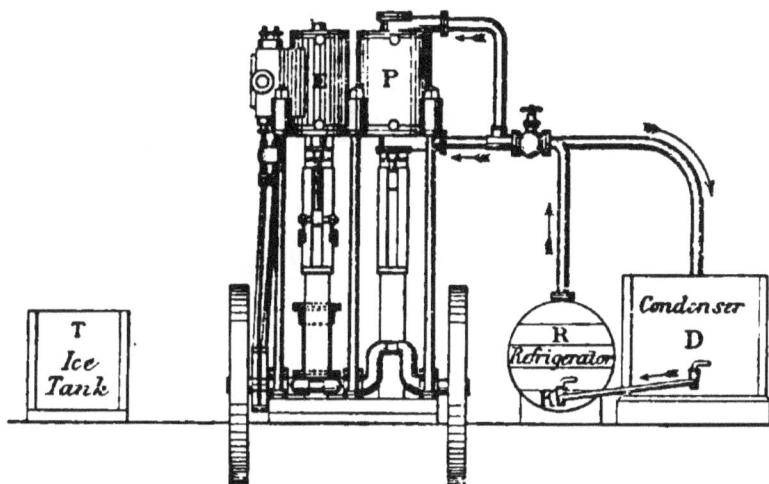

Fig. 5.

a mixture of five parts of alcohol of 90 per cent. strength and nine parts of concentrated sulphuric acid.

Ether mixes with alcohol in all proportions, but it is only soluble in nine times its own volume of water; it is very inflammable, and burns with a luminous white flame. As its vapour, mixed with the air, forms an explosive mixture, great care must be taken when using it to prevent leakage, as it explodes with the utmost violence. From its great disposition to fly off in the form of vapour, it is always dangerous to pour it out near a flame. The vapour tensions of sulphuric ether are shown graphically in Fig. 1, and are also given in the Appendix, Table B.

Messrs. Siebe, Gorman, & Co.'s machine applied to the manufacturing of clear ice, is shown in elevation and plan in Figs. 5 and 6. It consists of a refrigerator R,

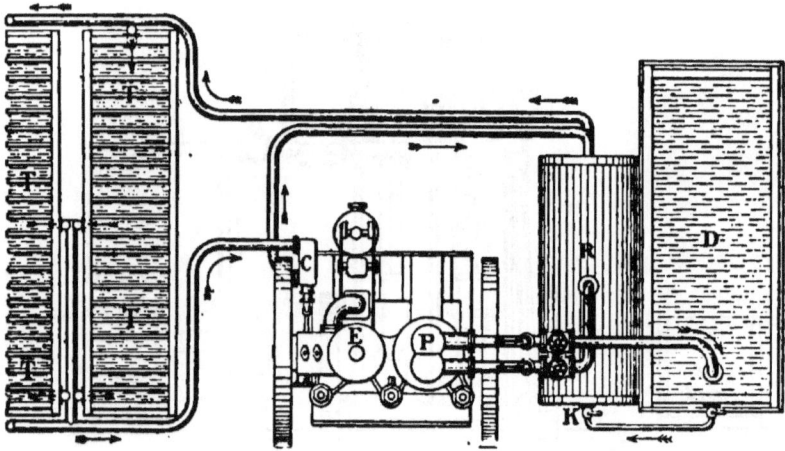

Fig. 6.

a water-jacketed pump P, driven by a surface condensing steam engine E, an ether condenser D, and ice-making tanks T, containing copper moulds, around which brine, cooled to a low temperature in the refrigerator, is circulated by a small pump C. The refrigerator R is a cylindrical vessel of sheet copper containing clusters of horizontal solid-drawn copper tubes, through which the brine successively circulates. The shell is connected with the large pump P by a pipe, the liquid ether from the condenser being admitted through a small pipe, having a cock K, which is so adjusted as to pass the precise weight of ether that the large pump P will draw off. What this weight is depends entirely on the pressure at which evaporation occurs; the greater the density of the vapour, the greater being the weight drawn off at each revolution of the pump. The pressure at which evaporation occurs is defined by the temperature to which it is desired to reduce the brine, the boiling point of the ether being regulated so as to give the required reduction of temperature and no more; otherwise the apparatus would not work up to its full capacity. The condenser D consists of a cluster of solid-drawn copper tubes, placed horizontally in a wood tank, through which cooling water is circulated, the amount of water required in this country being about 150 gallons per hour for every ton of ice made per twenty - four hours. With the temperature of cooling water available here, lique- faction generally occurs at a pressure of about 3 lbs. per square inch above the atmosphere; but in a warm climate the pressure needed may reach as much as 10 or even 12 lbs.

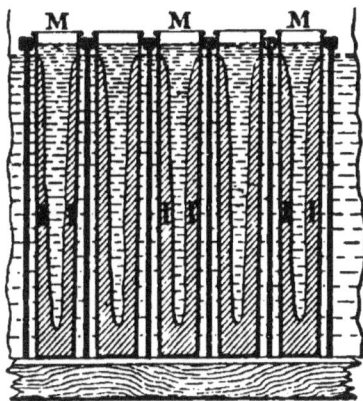

Fig. 7.

In this apparatus the ice is made in cans or moulds, as shown in section in Fig. 7. The moulds M of sheet copper or steel are filled with the water to be frozen and are suspended in a tank, through which is kept up a circulation of cold brine from the refrigerator. As soon as the ice I is formed, the moulds are removed and dipped for a few minutes in warm water to loosen the ice, which is then turned out. The sizes of the moulds vary a good deal, according to the capacity of the machine and the purpose for which the ice is to be used. A common plan is to commence with a thick-

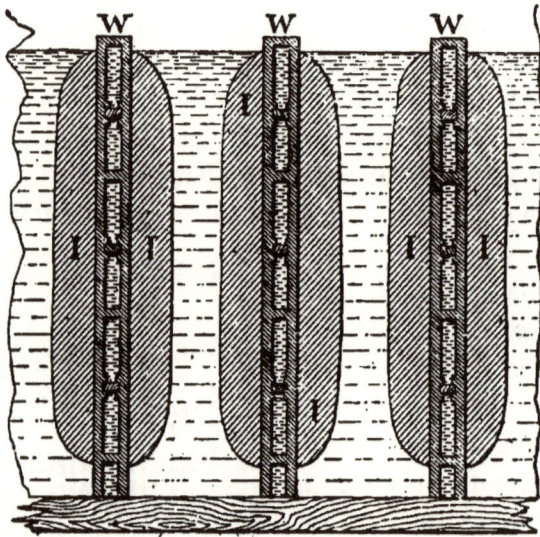

Fig. 8.

ness of 3 inches for a production of 1 ton per twenty-four hours, and to go up to 9 inches for 10 tons and upwards. The thickness exercises an important bearing upon the number of moulds to be employed for any given output; for, while a 3-inch block can be frozen in eight hours, one 9 inches thick will take about thirty-six hours. The time, however, varies also according to the temperature at which the brine is worked. For an ether machine such as that described, the brine temperature may be taken at from 10 deg. to 15 deg. Fahr.

Another method (shown in section in Fig. 8) is that known as the cell system; this consists of a series of cellular walls W of wrought or cast iron, placed from 12 to 16 inches apart, the space between each pair of walls being filled with the water to be frozen. The cooled brine circulates through the cells, the ice gradually forming outside, and increasing in thickness until the two opposite layers meet and join together. If thinner blocks are required, freezing may be stopped at any time, and the ice removed. In order to detach the ice from the walls, it may either be left for a time after the circulation of the brine has been stopped, or a quicker and better plan is to pass some warmer brine through the cells.

In order to produce transparent ice, it is necessary that the water should be agitated during freezing, so as to allow the escape of the air set free. When moulds are used (Fig. 7), this is generally done by means of arms having a vertical or horizontal movement, which are either pushed by the ice as it forms, leaving the block solid, or worked backwards and forwards in the centre of the mould, dividing the blocks vertically into two equal pieces. With cells (Fig. 8), agitation is generally effected from the bottom by means of paddles The ice which forms first on the sides of the moulds or cells is generally transparent enough even without agitation. The opacity gradually increases towards the centre, where the two layers meet and join together; agitation is therefore more necessary towards the end of the freezing process than at the beginning. As the quantity of air held in solution by water decreases as its temperature is raised, it is obvious that less agitation will be required in hot than in temperate climates; for this reason, in India and elsewhere, agitation is frequently dispensed with altogether.

Machines using ether as the refrigerating medium are also largely made by Messrs. Siddeley & Co., of Liverpool, and Messrs. West & Co., of Southwark; but they present no special features which are not embodied in the apparatus already described, the points of difference being in details to which it is not necessary to refer.

As already stated, the working pressure in the refrigerator must depend upon the extent of the reduction in temperature desired, bearing in mind that the higher the pressure the greater will be the work that can be got out of any given capacity of pumps. The liquefying pressure in the condenser depends on the temperature of the cooling water, and on the quantity that is passed through in relation to the quantity of heat carried away; this pressure determines the mechanical work to be expended. In any given machine the work may be accounted for as follows :—

Friction.

Heat rejected during compression and discharge.

Heat acquired by the refrigerating agent in passing through the pump.

Work expended in discharging the compressed vapour from the pump.

Against which must be set the useful mechanical work performed by the vapour entering the pump.

Assuming that vapour alone enters the pump, the heat rejected in the condenser is:

Heat of vaporisation acquired in the refrigerator, with the correction necessary for difference of pressure.

Heat acquired in the pumps, less the amount due to the difference between the temperature at which liquefaction occurs and that at which the vapour entered the pump, and less also the amount lost by radiation and conduction between the pumps and the condenser.

Though circumstances vary so much that no absolutely definite statement can be made as to the working of ether machines in general, the following particulars, taken from actual experiment in this country, will serve to show what may be expected under ordinary conditions :—

Production of ice per twenty-four hours	15 tons.
Production of ice per hour . .	1400 lbs.
Heat abstracted in ice-making per hour	245,000 units.
Indicated horse power in steam cylinder required for circulating the cooling water, and for working cranes, etc. . .	83 IHP.
Indicated horse power in ether pump	46½ IHP.
Thermal equivalent of work in ether pump per hour . .	119,261 units.
Ratio of work in pump to work in ice-making	1 to 2·05.
Temperature of water entering condenser	52 deg. Fahr.

4

For cooling water and other liquids, a similar machine to Messrs. Siebe, Gorman, & Co.'s, already described, is used; but in this case the ice boxes are dispensed with, the liquid being passed direct through the refrigerator without the employment of brine.

Methylic ether, a liquid with a latent heat of vaporisation of 473, and which boils under atmospheric pressure at 10·5 deg. below zero Fahr., has been employed by Tellier in apparatus of practically similar design to that used with ordinary ether. Its curve of vapour tensions is shown in Fig. 1. Tellier's apparatus has never come into use in this country, and need not be further dwelt on, for beyond the difference in size of pump, and the obvious alterations due to the higher working pressures, it presents no features of importance not possessed by the Siebe, Gorman, & Co.'s machine (Figs. 5 and 6).

In 1876, Raoul Pictet, of Geneva, successfully introduced sulphur dioxide as a refrigerating agent; and in France a large number of his machines have been made and put to work. In this country, also, they have been used, but to a much smaller extent.

Sulphur dioxide or sulphurous anhydride is a colourless gas of pungent, suffocating odour, with the well-known smell of the burning brimstone match. When breathed, it excites coughing, but is not dangerous unless it forms a large proportion of the atmosphere. It neither burns nor supports combustion. It is liquefied under pressure, has a latent heat of vaporisation of 182 deg., and under atmospheric pressure boils at 14 deg. Fahr. The vaporisation of the liquid acid causes the thermometer to descend to 68 deg. below zero Fahr. In the gaseous state its density is 2·247. Water dissolves fifty times its volume.

This body is prepared either by burning sulphur in oxygen, or by heating the metalloid with an oxygen compound of slight stability such as binoxide of manganese, or by deoxidising sulphuric acid by means of mercury or copper, and heat.

The vapour tensions of sulphur dioxide are shown in Fig. 1.

This machine is also of similar design to those in which ether is employed; but Pictet combined the refrigerator with the ice-making tanks, the brine being circulated by means of a fan. In this way the space occupied was reduced, and the efficiency somewhat increased. The cost of producing ice by the Tellier and Pictet machines may be taken at practically the same as that by the ether process.

An improved system of refrigeration by means of carbon dioxide or carbonic anhydride has lately been introduced by Messrs. J. & E. Hall Limited, of Dartford, Kent. It was first brought out in Germany, where its success led the present introducers to develop it in this country. Carbon dioxide, or carbonic anhydride, or carbonic acid, as it is variously termed, is gaseous at ordinary temperatures and under ordinary pressures. It is liquefied by a pressure of thirty-six atmospheres. When liquid carbon dioxide is thrown into the atmosphere, a part immediately vaporises, and absorbs in consequence such a quantity of heat that another part passes into the solid state. To collect the latter, the liquid jet is directed into a hemisphere of iron. The carbon dioxide may be kept solid for some time without its returning to the gaseous state. Pressed between the fingers, it destroys the skin like a hot body. When mixed with ether and placed in the receiver of an air pump, the temperature is lowered to -148 deg. Fahr. Gaseous carbon dioxide has a density of 1·529. It may be poured out in the air like a liquid.

Water dissolves its own volume of this gas, but does not combine with it to form an acid. Carbon-dioxide gas does not burn, and it neither supports combustion nor respiration. The principle upon which the machine works is by the evaporation of carbon dioxide, which is constantly recompressed and liquefied for further evaporation. The compressed gas is delivered into a condenser, consisting of coils of tubes kept cool by circulating water. The evaporation is produced by allowing the liquid carbonic anhydride to enter coils which are continually being exhausted by the compressor pump. Thus, the pressure being insufficient to maintain the gas in a liquid state, it evaporates at a low temperature, cooling the brine surrounding these coils. The brine, thus reduced to any desirable low temperature, is circulated by means of pumps through pipes in cold storage chambers, ice-making tanks, or is used for any other refrigerating purpose, the brine absorbing the heat from the surrounding objects.

The illustration (Fig. 9) represents a pair of J. & E. Hall's patent carbon-dioxide refrigerating machines, fitted with compound steam cylinders arranged side by side, having a surface steam condenser between them. The compressors are placed in line with the steam cylinders, the cranks of which are at right angles, by which arrangement an even turning moment is obtained.

Fig. 9.

Each compressor delivers its gas to an independent condenser, contained in the base of the machine, in which the carbon dioxide is condensed in the liquid form. In connection with each condenser there is a separate refrigerator, consisting of coils of wrought-iron pipes, in which the liquid carbon dioxide evaporates, producing intense cold. These coils are contained in a steel casing in which the brine is circulated.

Carbon dioxide, unlike ammonia, has no affinity for copper, hence that material can be used to resist the action of the sea water, which rapidly corrodes wrought iron, a point of the greatest importance when valuable cargoes are intrusted to the machine.

The machine illustrated, when running at its normal speed of eighty revolutions per minute, indicates 70 HP. The suction and delivery valves are made of tool steel, the valve being coned at 45 deg., whilst the seats, also of steel, are rounded, so that the bearing surface is very narrow. In this machine, even the neglect of the attendant appears to be provided for. Thus, in the event of any other machine being started without the delivery screw-down valve being previously opened, a breakdown having serious results might occur; but in this machine a safety valve has been fitted, preventing all possibility of accident.

The joints between the various parts of the refrigerating machine are absolutely tight, being made of any material suitable for the purpose, according to the working conditions. This is a feature with the carbon-dioxide machines, carbonic-acid gas attacking no metal or material, whereas, with ammonia and similar refrigerating agents, only iron or steel parts can be used.

The various parts of the machine are very much smaller than is the case with ammonia machines, whereby the strength is very greatly increased.

In the machines as illustrated, the two duplicate portions are so arranged that either the entire plant can be worked, or, if desired, one of the compressors, with its condenser and refrigerator, can be disconnected, when the other compressor can be worked by itself, the full advantage and economy of the compound engines being retained. Another point of advantage is that this duplex system gives all the security of two independent machines.

We understand that this type of machine has been designed to meet the limited space available on board ship, and to suit the usual height of the 'tween decks, machines even of the largest type, capable of preserving up to 60,000 frozen sheep, being less than 6 feet 6 inches in height.

It may be interesting to note a few facts with regard to carbon dioxide. Twenty years ago, the production of this liquid was a laboratory experiment; it is now sold at a few pence per pound, and large companies are formed in many parts of the world for its manufacture and sale. It is inodorous and non-poisonous in the quantity used. Carbonic acid is often confounded by the uninitiated with carbonic *oxide*, the latter being a most poisonous gas, and often causing loss of life in the form of fumes from coke or charcoal fires. On the contrary, the respiration of air containing even a considerable quantity of carbonic acid due to such an unlikely cause as a serious leak, is perfectly harmless, leaving no deleterious effect upon the system. As a further proof of the harmlessness of this gas, it may be mentioned that it is the gas which is used for making all aerated waters, and constitutes the "sparkle" of wines and bottled beers. To give an idea of the freedom from danger of this class of machine, it may be stated that the entire contents of the machine might be allowed to escape into an ordinary engine room without any of the disastrous results which would follow a similar escape of ammonia, sulphurous acid, ether, etc., used in other machines. We are informed that this experiment has been repeatedly tried.

Of course a small quantity is required to be added to the charge from time to time, to replace losses, but this is most easily done, even in the remote parts of the world, as the carbon dioxide is sent to any part of the world, in which it cannot already be obtained, in steel cylinders. To give an idea of the cost of first charging a twenty-four-ton ice plant with the gas, we understand that about £7 would cover it.

We understand that, besides ammonia machines, the Linde British Refrigeration Company manufacture special machines on the carbonic-anhydride system. This system was in fact originally introduced by Professor Linde, the first machine of this kind having been made by him for F. Krupp at Essen as far back as 1884. Since then several other machines have been made.

CHAPTER III

SOME of the more volatile derivatives of coal tar have been used in compression machines, especially in the United States; but it will be unnecessary to refer to them in detail, as their application has been exceedingly limited.

Anhydrous ammonia, that is, ammonia free from water, which may now be obtained as an article of commerce, has of late years been very largely introduced as a refrigerating agent, more especially in Germany and the United States.

Ammonia is composed of one part of nitrogen and three parts of hydrogen. It can be obtained from the air, from sal-ammoniac, and the nitrogenous constituents of plants and animals by process of distillation; as a matter of fact, there are very few substances free from it. At the present day almost all the sal-ammoniac and ammonia liquors are prepared from ammoniacal liquid, a bye product obtained in the manufacture of coal gas.

Pure ammonia liquid is colourless, having an extremely pungent odour and caustic taste. It turns red litmus, or test paper, blue. Its boiling point depends upon its purity, and is about 37 deg. Fahr. below zero at atmospheric pressure, and under this condition its latent heat of vaporisation, as determined by the highest authorities, does not exceed 600 thermal units, at which temperature one pound of liquid, evaporated under atmospheric pressure will occupy 21 cubic feet. Compared with water, the weight or specific gravity of the liquid at a temperature of 40 deg. Fahr. is 0·76, and the specific gravity of its vapour is 0·59, air being unity. The specific heat, or capacity for heat, of ammonia gas, as determined by Regnault, is 0·50836.

Ammonia is gaseous at the ordinary temperature and under the ordinary pressure; it liquefies at a low temperature or under great pressure; it may also be obtained as a white translucent solid fusible at −103 deg. Fahr. Its curve of vapour tensions is shown in the diagram (Fig. 1). Table C in the Appendix, which is compiled by Professor Wood, is compared with Regnault's celebrated experiments, showing the pressure of ammonia gas per square inch at the given temperatures Fahrenheit.

The simplest form of refrigerating apparatus employed consists of an evaporator in which the volatile liquid is vaporised. A pump which draws the gas or vapour from the evaporator as fast as formed. A condenser into which the gas is discharged by the pump, and under the combined action of the pump pressure and cold condenser the vapour is here reconverted into a liquid, to be again used in the evaporator.

It would appear from the foregoing that the pump and condenser might be dispensed with, and refrigerating effected by the simple apparatus shown in Fig 10, or

Fig. 10.

the slightly more complex form as in Fig. 11, where by means of the coils a greater cooling surface is exposed to the brine, resulting in a corresponding increase of efficiency. These conditions, however, may only be economically realised when the at present expensive

ammonia liquid can be obtained in great quantity, and
at less cost than the process of reconverting the vapour
into a liquid by the compressor pump and condenser.

The real index of the amount of cooling work possible
is the number of pounds of ammonia evaporated between
the observed range of temperature. To render this more
apparent, we may add that each pound of ammonia during evaporation is capable of storing up a certain quantity

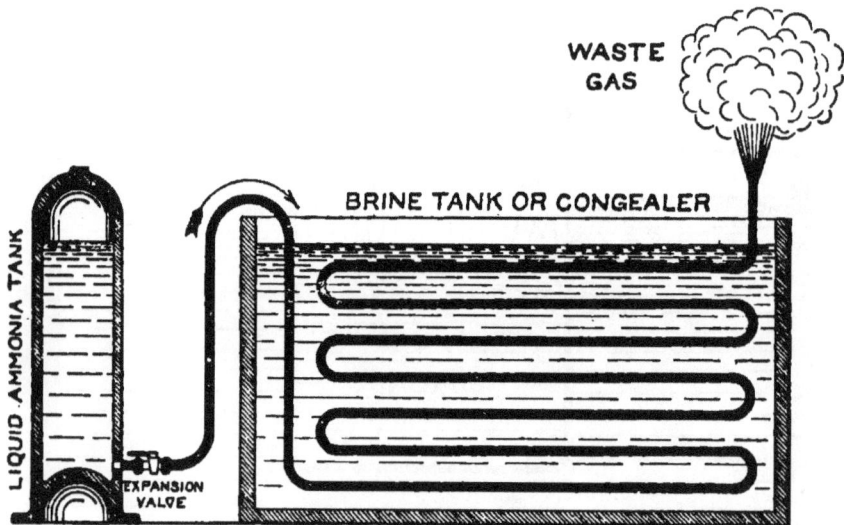

Fig. 11.

of heat, and that the simplest forms of refrigerating
apparatus might consist , as shown by Fig. 11, of two
parts, namely, an evaporator or congealer and a tank of
ammonia.   In this apparatus the ammonia is allowed to
escape from the tank into the congealer as fast as the
coils therein are capable of evaporating the liquid into a
gas ; when completely expanded, the resulting vapour is
allowed to escape into the atmosphere, which means it is

wasted, the supply being maintained by furnishing fresh tanks of ammonia as fast as the contents are exhausted. This process, while simple, would be tremendously expensive, costing, at present prices, about £50 per ton refrigerating or ice-melting capacity. To recover this gas and reconvert it to a liquid on the spot, in a comparatively inexpensive manner, is therefore the function of the compressor pump and condenser.

As previously mentioned, a liquid passing into a gaseous state, or when converted into a vapour, carries away a definite amount of heat from objects by which it is surrounded, and its capacity for storing heat under these conditions is called its latent heat of evaporation.

One pound of ice requires the application of 142 units of heat, and in round numbers the evaporation of liquid ammonia is equivalent to the cooling work done by the melting of four pounds of ice into water at 32 deg.

As ammonia will boil or evaporate, when exposed to the atmosphere, at a temperature of 37 deg. Fahr. below zero, or 249 deg. lower than the boiling point of water under the same conditions, it should be quite clear that a vessel of liquid ammonia thrust into a heap of snow at 32 deg., which is 69 deg. above the boiling point of ammonia, would bear about the same relationship to the snow heap as a vessel of water placed upon a fire. In both cases there would be an evaporation of the liquids and absorption of heat by the resulting vapour. To render this still more intelligible, we may state further, that the heat required to evaporate the ammonia is taken from the snow heap, which is made even colder than before.

Water similarly thrust into a snow heap would not boil, because the temperature of snow is far below its boiling point. Ammonia, on the other hand, would evaporate, because the temperature of the snow heap is above the boiling point of the liquid.

The somewhat popular idea of heat is that it should be hot enough to burn, and it is difficult for some persons to form any other conception. When it is clearly understood that the absolute or real zero on the scale is 461 deg. below the zero of the thermometers, it may readily be seen that within this great range it is more a question of the relative difference of temperature between two bodies brought in contact, that determines the amount of heat that is lost by the one and gained by the other, than the exact position in degrees they occupy upon the thermometric scale. The hottest will invariably impart its heat to the coldest until temperatures are equalised, and in the case of the ammonia, whose boiling point is 37 deg. below zero, it will continue to boil at atmospheric pressure, and carry off heat so long as it is in contact with any substance hotter than itself, or above 37 deg. below zero, making that substance continually cooler by absorbing its heat, or at least until it has been reduced to a temperature corresponding to the pressure under which the ammonia gas is formed; this point reached, the ammonia will cease to evaporate, and remain in a liquid state.

All the so-called permanent gases, such as hydrogen, nitrogen, and others, including the compound air itself, are simply the vapours or combination of vapours or gases of some liquid whose boiling points are so low upon the thermometric scale that the heat of the earth is sufficient to maintain them in the gaseous state, and at no time so low that they will be condensed and restored to their liquid state. It is only necessary to call attention to the numerous experiments made by prominent physicists who have proved this theory by subjecting all known gases to artificial cold and pressure, and condensing them to a liquid. This is the method in common use for liquefying gases, that is, to subject them to constant pressure, and, while in this state, to rob them of their latent heat of vaporisation by allowing cold water to flow over the apparatus, or in the case of the more permanent gases, at the same time subjecting them while in the condenser to the most intense artificial cold that can be produced, and thus liquefying them. Ammonia gas, by the combined effect of constant compression or pressure, and flowing water on the liquefier having a temperature slightly lower than the boiling point of liquid at the given pressure, is easily liquefied in any desired quantity.

This may not seem altogether plain at first sight, but we will try and explain the reason. For example, it is well known that increasing the pressure upon a liquid raises its boiling point, and from this fact it can easily be imagined that liquid ammonia, which boils at 37 deg. below zero under the pressure of the atmosphere, would have its boiling point raised in proportion to the pressure to which it is subjected, and this may be carried to such an extent that the boiling point of ammonia can be thus mechanically raised to any desired temperature. It should also be noted that gas at a given pressure must have a temperature or boiling point to correspond. For instance, the boiling point of water at atmospheric pressure, 15 lbs.

absolute, is 212 deg., and the temperature of its gas or
steam is likewise 212 deg.

Pressure and temperature are interrelated.   That is to
say, that at a certain pressure saturated gas has a corresponding temperature, and this is also the case with
ammonia, see Table C.   Hence, if the gas, while subjected to a certain uniform pressure, be discharged into
an air-tight vessel, which is being constantly cooled by
water of a somewhat lower temperature than that due
to the pressure of the gas, the vapour will necessarily
under these conditions collapse and be condensed inside
the vessel, and go back to the liquid form.

The temperature of the water available for use on the
condenser determines the pressure to which the gas must
be subjected in order to raise its boiling point high
enough, so that it cannot exist as vapour when chilled by
contact with condensing surfaces slightly colder than
itself, but will collapse.   The condenser is an interesting
application of a natural law, made clear by investigating
the relations between boiling points and pressures.

It may here be stated that in the use of ammonia two
distinct systems are employed, namely, the compression
already described, and the absorption.   So far, however,
as the mere evaporating or refrigerating part of the
process is concerned, it is the same in both; the object
being to evaporate the liquid anhydrous ammonia at such
tension and in such quantity as will produce the required
cooling effect.   The actual tension under which this
evaporation should be effected in any particular case
depends entirely upon the temperature at which the
acquirement of heat is to take place, or, in other words,
on the temperature of the material to be cooled.   This

will be understood by reference to the diagram Fig. 1.
The higher the temperature, the higher may be the
evaporating pressure, and therefore the higher the density
of the vapour, the greater the weight of liquid evaporated
in a given time, and the greater the amount of heat
abstracted. On the other hand, it must be remembered
that, as in the case of water, the lower the temperature
of the evaporating liquid, the higher is the heat of
vaporisation. It is in the method of securing the rejection of heat during condensation of the vapour that the
two systems diverge, and it may be better to consider
each of them separately.

So far as the cycle of operations is concerned, it is
exceedingly simple, being precisely the same as for ether.
The apparatus being charged with a sufficient quantity of
pure ammonia liquid, which we will for simplicity
assume to be stored in the lower part of the condenser,
a small cock or expansion valve controlling a pipe
leading to the refrigerator or brine tank is slightly
opened, thus allowing the liquid to pass into the evaporator coils. These coils really perform the same office as
a tube or flue in steam boilers, and have precisely the
same function, and may be called the heating surface.
The amount of water capable of being boiled into steam
in a boiler depends upon the square feet of heating
surface, temperature of fire and pressure of steam ; and
the same is true of the capacity of heating surface presented by the coils in the evaporator. The heat is trans-
mitted through the coils from surrounding bodies, which
are thus reduced in temperature or refrigerated, the heat
being absorbed by the ammonia liquid which is boiled
into a vapour the same as water is boiled into steam in

5

an ordinary boiler. As previously explained, the surrounding substances part with an equivalent amount of heat, and thus become cooler, the amount taken up and made negative being in proportion to the pounds of liquid ammonia evaporated. The amount of mechanical cold is easily regulated by the cock or valve leading from the condenser. As the gas begins to form in the refrigerator, the pump is set in motion at such a speed as to carry away the gas as fast as formed, which is discharged into the condenser under such pressure as will bring about a condensation and restore the gas to the liquid state, the operation being continuous so long as the machinery is kept in motion. The ammonia thus recovered flows back into the refrigerator as required, and is there again evaporated, so that the small quantity of ammonia forming the charge of the machine is continually subjected to the same cycle of operations.

In the construction of ammonia as compared with ether machines, there are two essential points of difference. For, in the first place, the pressure of the ammonia vapour is much higher than that of ether at the same temperatures, its tension at 60 deg. Fahr. being 108 lbs. per square inch; and, secondly, owing to the action of ammonia on copper, no brass or gun metal can be used in any part with which the gas or liquid comes in contact. One of the chief difficulties encountered in the compression of ammonia is leakage at pump glands. The gas is extremely searching, and even at the comparatively low pressure of 30 lbs. per square inch above the atmosphere, it will readily find its way through an ordinary gland; while at the pressure existing in the condenser, which may be taken at from 150 to 180 lbs.

per square inch, this tendency is of course much aggravated. In order to minimise the leakage and to simplify the construction of the gland, the pumps are frequently made single-acting, as in this way the gland is exposed only to the refrigerator pressure, which is seldom above 30 lbs. It is also usual for glycerine, or some lubricant that does not saponify with ammonia, to be injected into the pump, so as to form a liquid seal for the gland, and in some cases for the piston as well; this is the general practice in the United States. In Germany, on the other hand, where the compression machine has been very largely applied, the double-acting pump is more usual. To lessen leakage, Linde provides a chamber in the gland box, into which glycerine or some suitable lubricant is constantly forced at a slightly greater pressure than that prevailing in the condenser, so that the tendency is for the lubricant to leak inwards, instead of ammonia outwards. Any lubricant that does get into the pump passes out with the ammonia, and is separated from it in a suitable vessel.

With regard to the other parts of the apparatus, but little need be said. Wrought-iron coils or zigzags are used for both the condenser and the refrigerator, their precise form depending on the fancy of the designer. The refrigerator is generally combined with the ice tanks, the cooling pipes being placed either below or at the side of the moulds, sometimes in a separate compartment, and sometimes in the same tank. With the cell system shown in Fig. 8 an independent refrigerator is used, the cold brine being circulated by a pump in a similar manner to that described for the ether system. Owing to the low temperature which may be attained by the use of am-

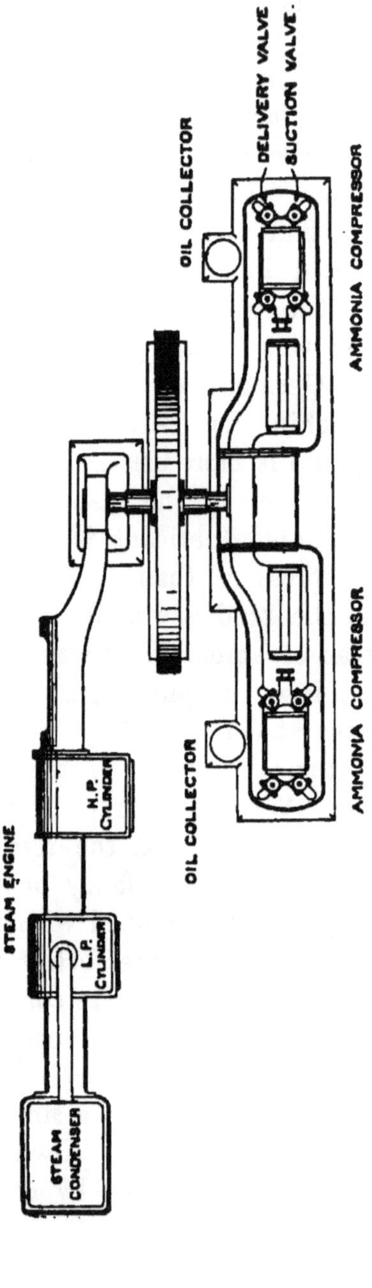

Fig. 12.

STEAM ENGINE

STEAM CONDENSER

L.P. CYLINDER

H.P. CYLINDER

OIL COLLECTOR

OIL COLLECTOR

AMMONIA COMPRESSOR

AMMONIA COMPRESSOR

DELIVERY VALVE

SUCTION VALVE

monia, care has to be taken in the selection of a brine that will not congeal with the degree of cold to which it will be subjected. A solution of calcium or magnesium chloride in water is generally used.

Fig. 12 is the plan of a land-type machine, with double ammonia compressors, driven by a tandem compound jet condensing engine, manufactured by the Linde British Refrigeration Company Limited. As the different parts of the machine are clearly marked, the reader, from the descriptions already given, should have no difficulty in understanding its mode of working. It may be mentioned that the condenser is not shown in the plan, as it can be placed in any convenient position, and any suitable condenser may be employed.

Figs. 13 and 14 show the elevation and plan of a marine - type single - compound ammonia - compression machine, driven by a tandem compound engine, manufactured by the same firm. The chief advantage of a compound compressor is that the compression of the ammonia is accomplished in two stages whereby the loss in clearance occurs in the LP compressor only, that in the HP being obviated entirely. The chief points to be remarked in these machines are that the coils in the condensers are endless, and that all joints are also endless.

The elevation and plan of a marine-type duplex-compound ammonia-compression machine, driven by a compound engine, are shown in Figs. 15 and 16, and are also manufactured by the above company.

These machines, which are of the latest designs and embody all the most recent improvements, need not be detailed further.

The mechanical work expended in compressing am-

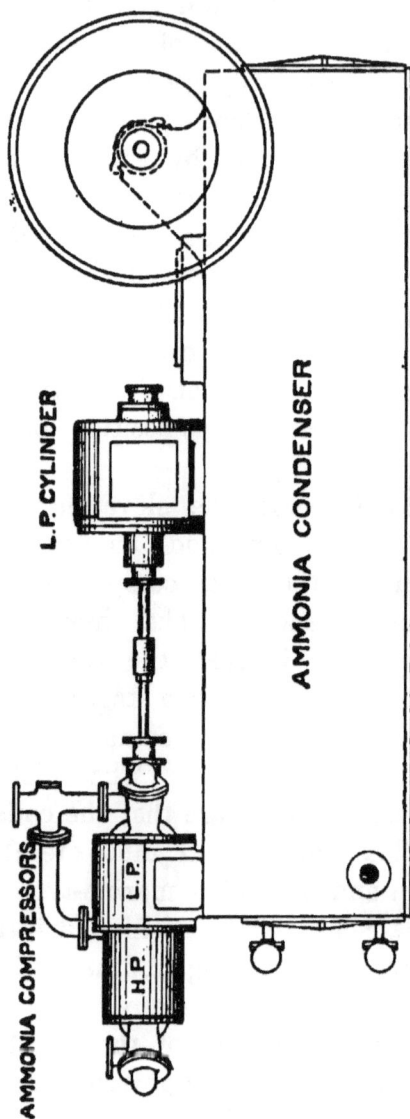

L.P. CYLINDER

AMMONIA CONDENSER

AMMONIA COMPRESSORS

L.P.

H.P.

Fig. 13.

monia may be accounted for in a precisely similar manner to that expended in the compression of ether. Notwithstanding that the degree of compression is so much greater with ammonia than with ether, the energy expended in compressing, heating, and delivering the gas is less, owing to the much smaller weight of ammonia required to produce a given refrigerating effect, the weight being in the inverse ratio of the heats of vaporisation, or as 1 to 5·45. For this reason the cost of making ice is much less with ammonia than with ether; one ton of coal being capable of producing as much as 12 tons of ice in well-constructed ammonia apparatus having a capacity of 15 tons per twenty-four hours. With coal at 15s. per ton, the cost of making ice by the ammonia-compression system may be taken at about 3s. 9d. per ton for a production of 15 tons per twenty-four hours, exclusive of allowances for repairs and depreciation.

Through the courtesy of the manager of the Linde British Ice Company, the author is enabled to give the following results of a test made by a committee of Bavarian engineers with a machine erected in a brewery in Germany. The test, he believes, was carried out in an impartial manner; and though it is not put forward by the Linde Company as showing the results attained in the ordinary working of their machines, it will nevertheless be of interest as indicating what may be expected under the most favourable conditions :—

Nominal capacity of machine, ice per twenty-four hours . . . . 24 tons.

Actual production of ice per twenty-four hours 39·2 tons.

Actual production of ice per hour . . 3659 lbs.

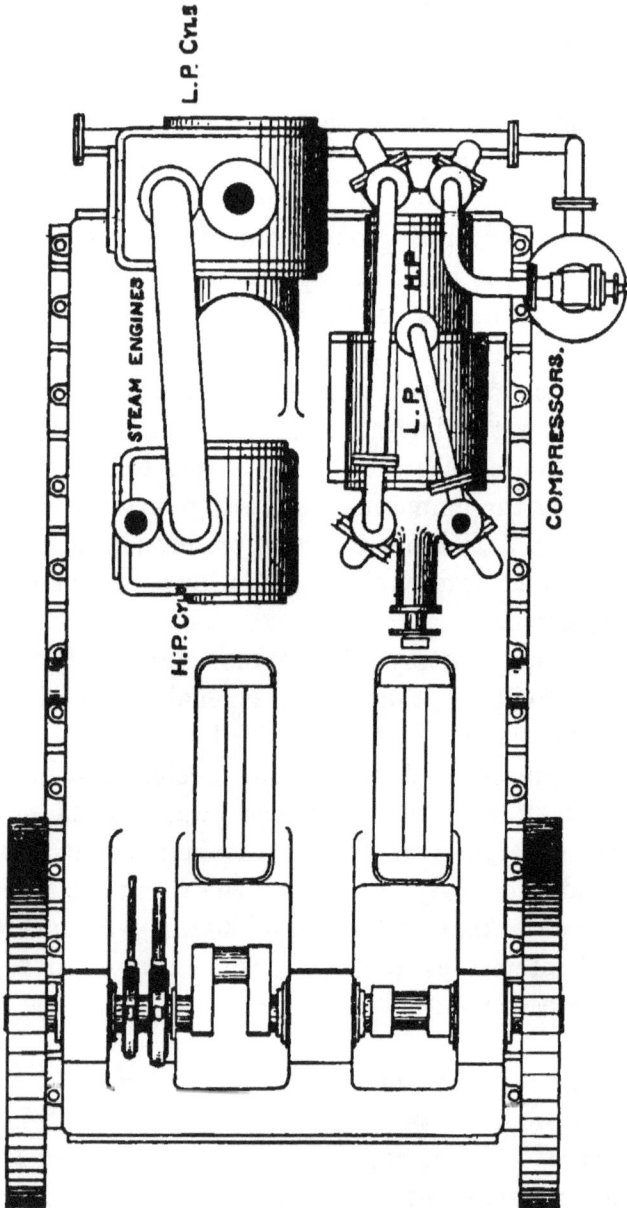

Fig. 14.

Heat abstracted in ice-making per hour, 731,800 units.

Indicated horse power in steam cylinder, excluding that required for circulating the cooling water and for working cranes, etc.	53 IHP.
Indicated horse power in ammonia pump	38 IHP.
Thermal equivalent of work in ammonia pump, per hour	97,460 units.
Ratio of work in pump to work in ice-making	1 to 7·5.
Total feed water used in boiler, per twenty-four hours	26,754 lbs.
Ratio of coal consumed to ice made, taking an evaporation of 8 lbs. of water per lb. of coal . . .	1 to 26·3.

In this case the pumps were driven by a Sulzer engine, which developed one indicated horse power with 21·8 lbs. of steam per hour, including the amount condensed in steam pipes.

Ammonia compression machines are also manufactured in the United States on a large scale by a number of firms, notably by the De La Vergne Refrigerating Machine Company of New York.

This company furnish machines ranging from a capacity of 3 cwt. of ice per day to 220 tons per day, and Fig. 17 is a side elevation of one of the latter machines.

A sectional view of one of their double-acting machines (Fig. 18) gives a very good idea of their general construction. It will be observed that the engine which supplies the motive power is horizontal, while the machine itself is vertical; this style is almost exclusively

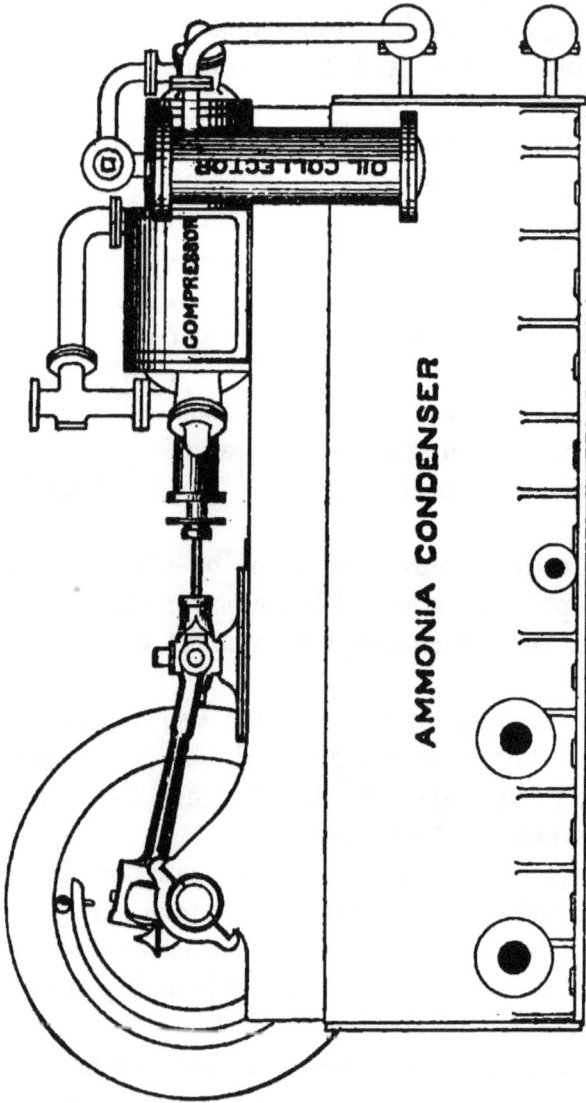

Fig. 15.

American, and seems a general favourite on the other side of the Atlantic.

To make mechanical refrigeration a success, it is essential—1, to discharge the entire volume of the gas entering the compressors; 2, to prevent all leakage past the stuffing box, piston, and valves; and 3, to extract the heat from the gas during compression. All this the De La Vergne Company claim to accomplish by a simple device, one for injecting into the compressor, at each stroke, a certain quantity of lubricating liquid, which effectually seals the stuffing box, piston, and valves, fills all clearances, and takes up the heat developed during compression.

For a number of years they have been experimenting to solve the problem of constructing a double-acting compressor which would handle the gas in connection with their system of oil circulation as well on the up *and* down stroke as the single-acting compressor does on the up stroke. It is apparent that a double-acting pump is more advantageous—providing it is well constructed—because it handles double the amount of gas with every revolution of the crank-shaft that a single-acting compressor does which has the same diameter and the same stroke. The moving parts, such as cross head, piston, piston rod, and connecting rod, being the same for either a single or a double-acting compressor, the *friction will be the same for all these parts, while double the work is being effected.* To overcome friction means power expended—*power wasted*—and in their case, viz. in a machine with two gas compressors, it means a saving of one-eighth of the whole power used for compressing the gas. Another advantage is the cheapening of the machine

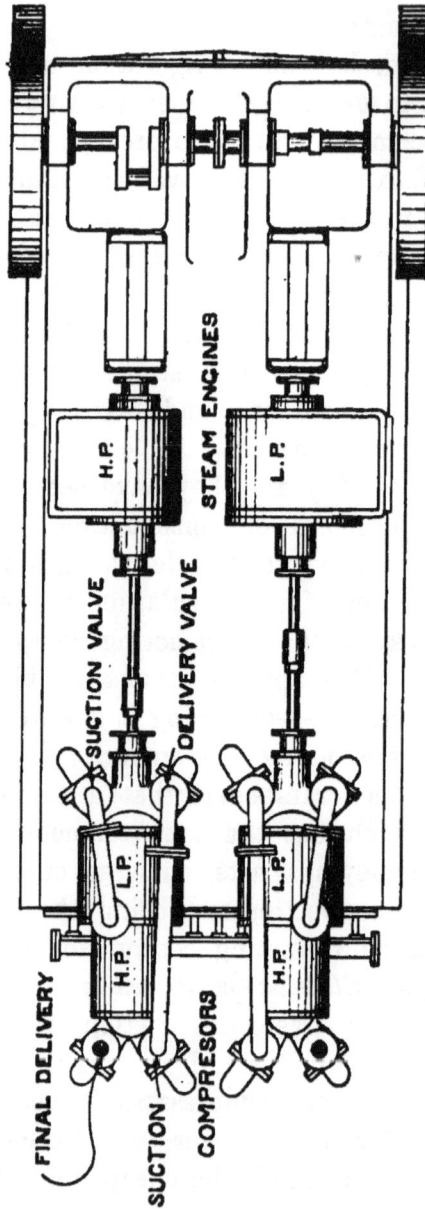

STEAM ENGINES

H.P.

L.P.

SUCTION VALVE

DELIVERY VALVE

L.P.

L.P.

H.P.

H.P.

FINAL DELIVERY

SUCTION

COMPRESORS

Fig. 16.

through the fact that one double-acting compressor will do the work of two single-acting ones of the same size.

In the ordinary form of double-acting compressors the discharge valves at the lower end are placed either on the side or in the lower head. In either case the oil is discharged on the down stroke *before* all the gas has left the pump—and this is wrong. The oil must be discharged *after* all the gas is gone, because otherwise re-expansion takes place, and this means loss of efficiency of the pump. They have avoided this difficulty in the following manner:—

At the lower end of the compressor (Fig. 19) there are two discharge valves placed on the side—one above the other. On the down stroke either of the valves or both may open until the piston covers the upper one, when only the lower one is open to the condenser. In the further course of the piston and as soon as the lower valve is also closed, the upper one is in communication with an annular chamber contained in the piston. This chamber has valves in its bottom which open into it as soon as all other outlets from the lower side of the piston are closed (they open a little harder than the discharge valves on the side), and which allows all the gas to go out through the piston, and, after the gas, the oil follows, thus permitting no gas to remain on the lower side after the completion of the down stroke. It will be seen that in this manner the very important oil system of their machine is retained, and that the lower side of the pump works as well as the upper, while the oil effectually seals the stuffing box in spite of the higher pressure on it at the end of the down stroke.

The machines with this style of compressor have been

Fig. 17.

220 TON REFRIGERATING MACHINE.

in operation, some of them, nearly five years, and have all given the utmost satisfaction, and the De La Vergne Company are now recommending them as superior to the single-acting machines on account of the saving in power and greater cheapness.

This company state that they prefer to refrigerate establishments by expanding the gas direct through pipes placed in the room to be cooled, and not by first cooling a non-congealable salt brine and pumping this through pipes in the room. The reason for this is that a loss of efficiency is always connected with every transmission of heat. They can carry an evaporating pressure in their direct pipes of 25 lbs. and still have within them a temperature of 14 deg. Fahr., while only 15 lbs. can be carried in the evaporating coils to keep the brine at 18 deg. The result is that the gas is forced into the compressors at a higher back pressure than if the brine were first cooled; and, as explained before in enumerating the advantages of ammonia over other agents, they get a greater efficiency from a certain compressor the greater the pressure is at which it takes in the gas. Furthermore, the cold is produced just where it is wanted, and nothing is lost, while in the brine system a large tank is exposed to the atmosphere, and, even if insulated, absorbs a great deal of heat, which is a total loss. To pump the sometimes immense masses of brine through thousands of feet of pipes, which after a while become coated on the inside with rust and slime, and thereby produce great friction and non-conductibility, costs a considerable amount of steam, so that, through all these different causes combined, it is claimed that the efficiency of their machines is considerably increased.

The objection raised against placing the ammonia pipes direct into the rooms to be.cooled, that there is

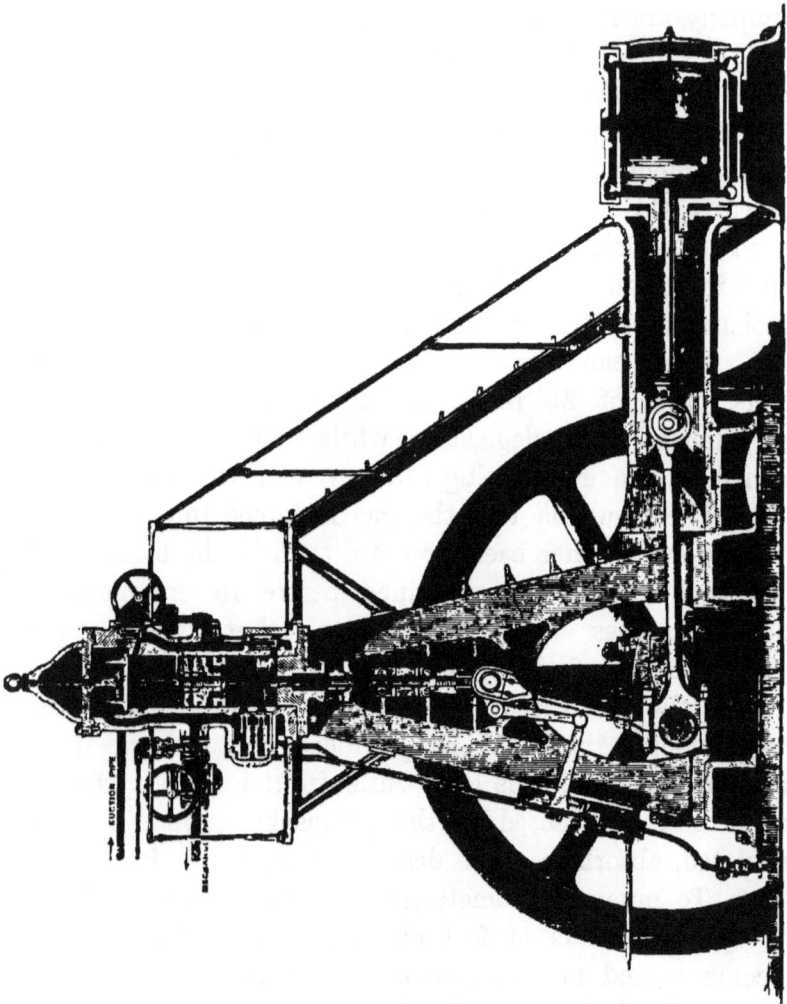

Fig. 18.

danger of leakage, is met by a system of pipe connections and cocks.   All pipes are tested singly, before they are put up, to 1000 lbs. hydrostatic pressure per square inch; and after they are all connected, the whole system is subjected to an air pressure of 300 lbs., at which the gauge must remain for hours in succession. In this manner the company produce a plant that is many times safer than any steam boiler; and since the first machine was erected, in 1879, and with over 1000 miles of pipes now in operation, there is not a single accident yet to record. The size of pipe adopted by them for the expansion coil is 2 inches diameter, and preference is given to this size for the reason that it is lap-welded, whereas the smaller sizes are butt-welded; and also on account of the diminished friction of the gas in passing through pipes of this diameter.

Formerly the company used pipes *only* to obtain the necessary cooling surface in the rooms to be refrigerated; but since 1882 they have accomplished the same object by means of cast-iron discs, which are made in halves and attached to the expansion coils, after these are all put up, by means of iron clips, which press the two halves together against the pipes. The cooling surface is thereby increased to such an extent that, where formerly it required four, only one foot of pipe is now needed, thus saving in room and first cost.

The application of the disc is based upon the principle now used in the most efficient of our modern steam radiators, in which the heating surface exposed to the air is increased by means of flanges and projections added to the outside surface of the radiator; thus exposing a larger heating surface than was attained with the old form of steam coils.

By applying these discs to steam coils the same results could be obtained as with the modern steam

6

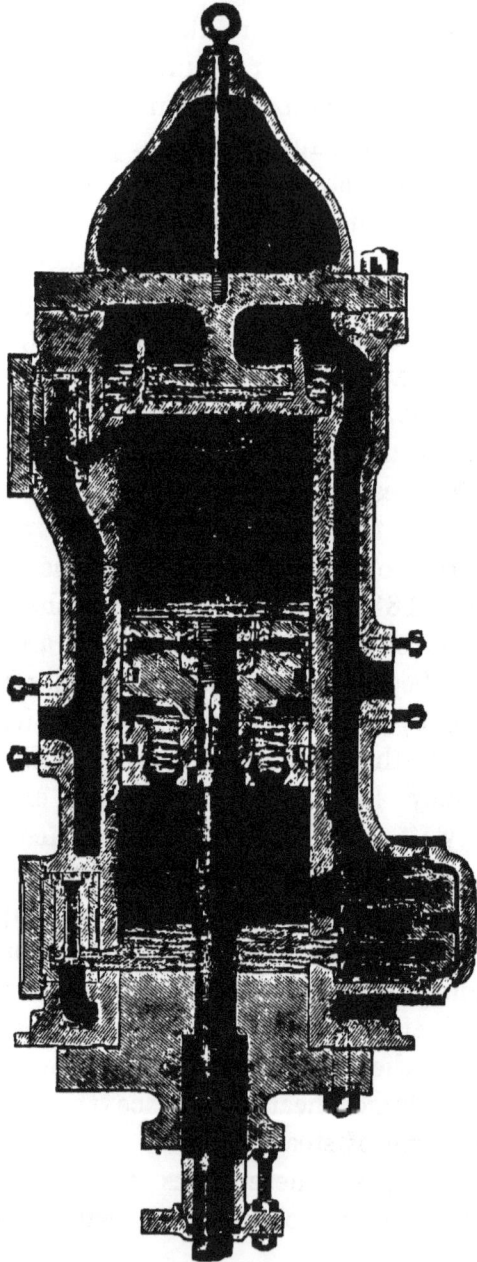

Fig 19.

radiator; the transmission of heat could be increased or diminished according to the number of discs applied to each lineal foot of pipe.

The results obtained are based upon the fact that heat is conducted with more rapidity by iron than by air. Whereas, one square inch of iron will transmit, say, 50 heat units per minute to another piece of iron attached to its surface, it will transmit but one heat unit, under similar conditions of temperature to air.

In order to make a refrigerating coil quick and effective in reducing the temperature of the air, the air is brought in contact with as large a refrigerating surface as practice admits of, without, however, increasing the internal surface bathed with the chilled liquefied ammonia to more than is absolutely necessary.

It may be as well to state here that ammonia has no chemical effect upon iron; a tank, pipe, or stop cock containing ammonia in a gaseous or liquefied condition will stand an indefinite time, and, upon opening, no action will be apparent. Pipes have been in use for twelve years, the inside surfaces of which have not changed one particle. The only protection, therefore, that ammonia-expanding pipes require is from corrosion on the outer surface. As long as the pipes are covered with snow or ice corrosion does not occur; the coating of ice thoroughly protects them from the oxidising effect of the atmosphere; but alternate freezing and thawing requires protected surfaces, which are best obtained by applying a coat of paint every season.

A great drawback in the construction of refrigeration machines has been the difficulty of making a pipe system with its joints and cocks, or valves, which was *absolutely tight*, so that there would be no obstacle in the way of using as many joints and cocks as was found necessary for a perfect system of direct expansion.

To ensure perfect tightness between the pipes proper

Fig. 20.

and the fittings to which they are attached, the De La Vergne Company have invented and patented a joint which is called the " screwed and soldered " joint. In the selection of fittings, which are shown on Figs. 20 and 21, it will be seen that the thread into which the pipe screws does not reach entirely to the outside. It is enlarged to the depth of $\frac{1}{2}$ to $\frac{3}{4}$ of an inch, forming a smooth annular space around the pipe beyond the termination of its thread. All the fittings are made of malleable iron or steel, which admit of being well tinned, and thus form a screwed and soldered joint by entirely filling the annular recess, formed on the outside by the

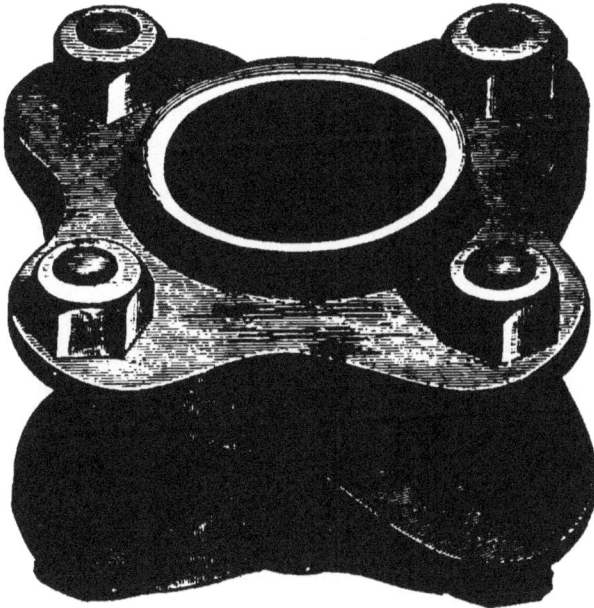

Fig. 21.

fitting, and on the inside by the pipe, with solder. The result is that the thread of the pipe is entirely covered, and that the otherwise weakest part of the pipe is made the strongest. In overrunning the test pressure of 1000 lbs. to the square inch, at which all their pipes, fittings, and cocks are tested to the point of bursting, the pipe is always ripped open before this joint gives out.

The De La Vergne machines are manufactured in this country by Messrs. Sterne & Company Limited, of London and Glasgow, who, we understand, are the sole licensees for Great Britain.

The "Eclipse" refrigerating and ice-making machines, which are perhaps better known in the United States than in this country, are manufactured by the Frick Company, of Waynesboro, Pennsylvania, to whom we are indebted for the following particulars :—

One of their standard machines is shown in Figs. 22 and 23, its capacity being 150 tons. The type of machine is seen by the engravings to consist of a pair of single-acting vertical compressor pumps driven by a horizontal steam engine, all forming part of one structure.

The principal feature in the design, which is exclusively their own and protected by patents, is the employment of rectangular tapering box-girder columns for supporting the pumps. The bottom ends of the columns terminate in a broad base or flange, which finishes flush with the floor line and is bolted to heavy foundation girders. The upper works are provided with a gallery and convenient stairways. On the larger machines, an upper and lower platform are furnished. The flywheel is located between the pumps; the connecting rod of the horizontal engine acts directly upon the main crank through an opening in one of the columns. Altogether the design is novel and pleasing, and strikes the observer as being a splendid adaptation for the purpose to which the machine is applied. It will be noticed that the design affords easy access to all the working parts, an unusual advantage in this class of machinery.

The most important feature of a refrigerating machine

Fig. 22.

is the compressor pump. To secure the highest efficiency
of performance (other things being equal, such as proper
application and proportion of the steam engine driving
the same, with the lowest obtainable loss of friction in
transmission of power to the pump), the pump which
receives the fullest charge of gas and most perfectly
expels the same is the most efficient *and will do the
most work*.

Fig. 24 is a sectional view of the " Eclipse " gas-com-
pressor pump No. 1, showing self-contained and removable
valve mechanism, with safety head and independent or
false seat and cone bonnet separate from pump casting.
An improved form of compressor pump is shown in Fig.
25, it being fitted with a self-contained patent safety
head and improved discharge valve. In this construction
it is simply necessary to remove the light pump head, to
expose the valve mechanism, which can then be bodily
lifted out for examination ; or in case of repair, a
duplicate set inserted, no other joint than the pump
head being disturbed. The base of the pump containing
stuffing box and lubricating mechanism is similar to
pump shown in Fig. 24.

The pump piston is gas-tight, by reason of its excellent
fit and extra length, yet moving freely, compensation for
wear being provided by using five spring rings, which are
carefully fitted.

Particular pains are taken with the pump bore to have
it round, parallel, and smooth.

The steel suction valve, of large area, is situated in the
piston, the gas inlet being in the base of the pump. The
suction valve being balanced by a spring, presents upon
the return stroke of the piston no resistance to the gas,

Fig. 23.

which flows, under the back pressure, with considerable velocity into the vacant space above the piston, *hence no check valve is needed in suction pipe in order to fully charge the pump.* A cushion and spring assist in closing the suction valve promptly and noiselessly as the up stroke is begun, the imprisoned gas being gradually compressed until it equals the condensing pressure acting upon the discharge valves, located in the pump dome, when discharge begins.

The gas, under the treatment it receives in the compressor pump, parts with more or less of the heat of compression through the wall of the cylinder, the dome and cylinder being enveloped by a water jacket, through which the cold water is constantly circulating.

This jacket water not only prevents superheating of the gas during compression, but besides carries off much of the heat, also materially assisting the condenser and cutting down to a marked degree the gas resistances that would present themselves without its use. As a matter of fact, it is so much gained at the slight cost of the jacket water.

In order to make it perfectly safe to work the piston *metal to metal* against the top cylinder head, the better to expel the full charge without danger, the pump head is made movable, or what may be called a *safety head*; in other words, it is simply a large valve, the full size of bore of pump, through the seat of which the piston may pass without injury, raising the head before it sufficiently, in case of any part getting loose, that no damage can ensue, such as knocking out a cylinder head, which in some constructions is frequent, thus losing the full charge of ammonia gas and endangering life.

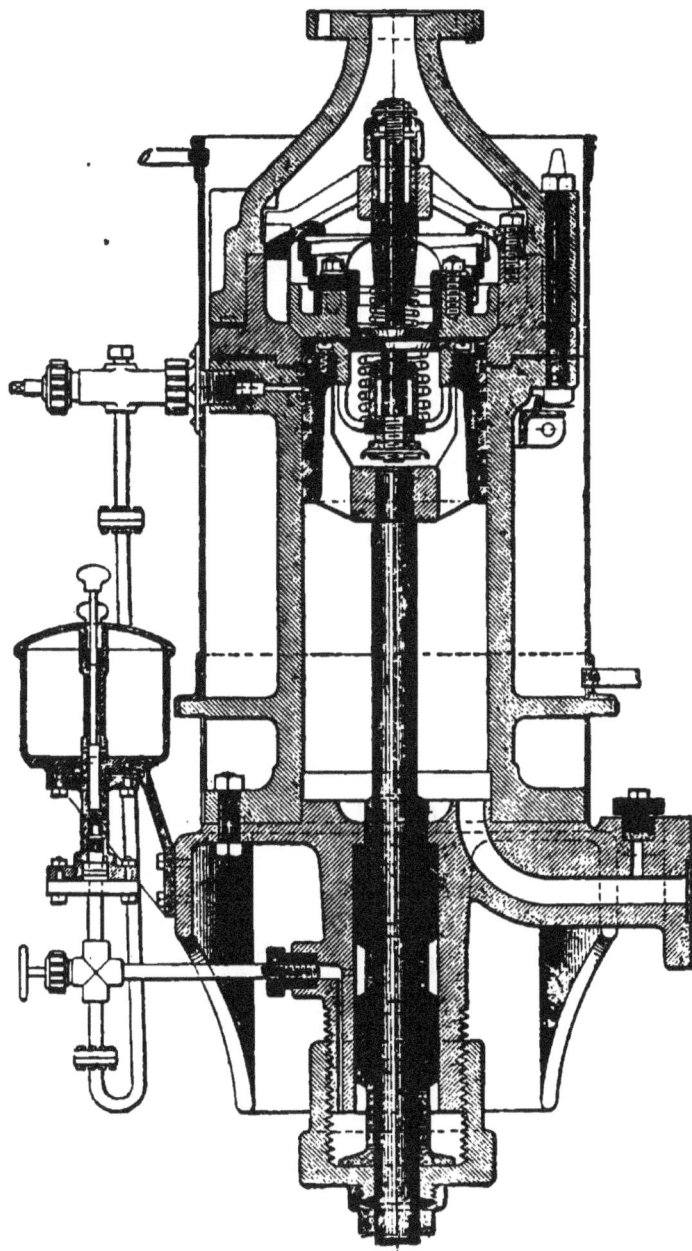

Fig. 24.

The safety head does not work as a valve, the real operating discharge valve being the small steel valve in the centre of the same. Notice that the safety head with its discharge valve, guides, and seat are *self-contained and independent* of the pump cylinder, making it convenient to replace the whole valve mechanism by a duplicate one or make speedy repairs.

The stuffing box for the piston rod is arranged with great care to prevent leakage of gas, and amply lubricated by means of a hand pump and oil reservoir.

The small valve at the upper left-hand corner of diagram is for taking indicator diagrams from the pump, and to inject oil in that portion of the cylinder, if any be needed, when starting up a new machine or when testing under air pressure.

A novel feature that will be appreciated when studying the design of the compressor is that *all parts of the pump mechanism are easy of access*; by simply removing the dome head, the valves are exposed to view for examination or adjustment, and the entire valve mechanism can be removed in a twinkling. The peculiar arrangement of the pump dome facilitates the removal of the parts without disturbing the main discharge pipes or stop valve, but one joint being broken.

It will have been noticed from Figs. 24 and 25 that the "Eclipse" pump is a single-acting vertical one, while a few other builders use double-acting pumps, some of them horizontal.

The reason for using this style is that vertical pumps are not subject to bottom wear of the pistons, as is the case with horizontal pumps, where the weight of the piston is necessarily supported by the cylinder bore, pro-

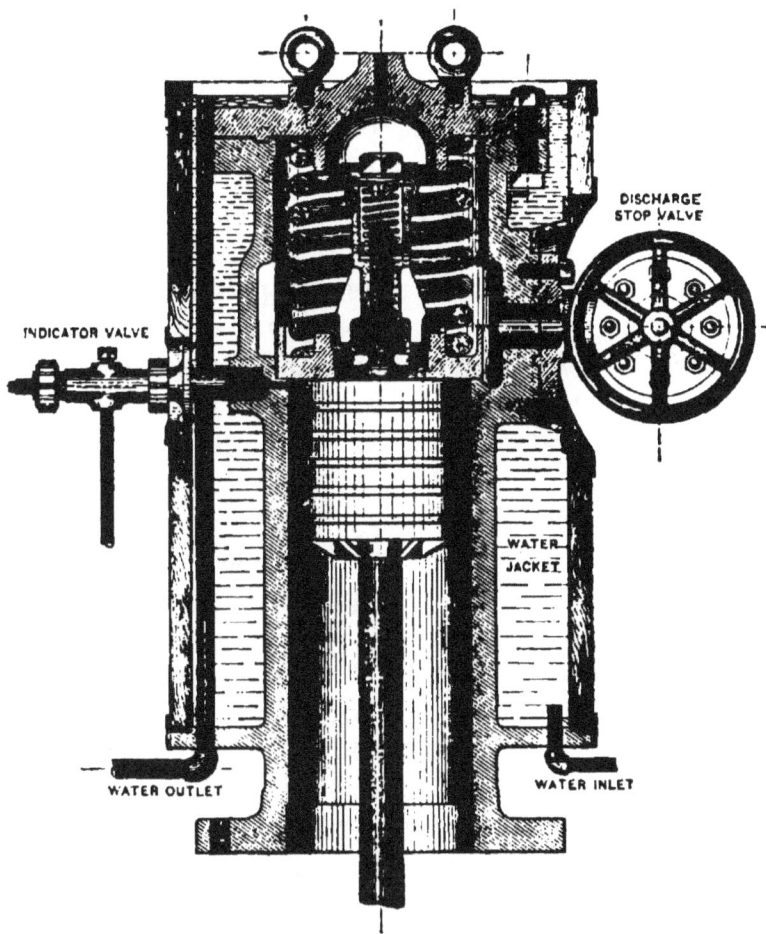

Fig. 25.

ducing needless friction, with a strong tendency to wear the cylinder oval, and more particularly the narrow surface of the piston to such an extent that leakage of gas occurs past the top of same. Part of the weight of the piston is also taken by the piston-rod when near the front end of cylinder, and rests on the stuffing box, which makes it rather difficult to maintain the piston rod packing.

In the vertical pattern the pump cranks are placed opposite each other and the forces are balanced. The wear of the pump cylinder bore and piston rod, owing to their vertical position, is uniform and very slight, and the saving in friction a considerable item. There are other advantages, such as economy of space, use of a large vertical water jacket, separation of dirt, ease of lubrication, etc.

The " Eclipse " single-acting pump *compresses the gas on its upward stroke*, hence the condensing pressure comes only above the piston. This pressure ranges from 125 lbs. upwards per square inch. The space below the piston being subjected only to the low suction pressure of gas, ranging from 0 to 35 lbs., in a single-acting machine, the stuffing box packing is more easily kept tight without undue wear and friction of the piston rod as in the double-acting pump, which, having to *compress gas on the lower stroke* to condensing pressure, necessitates a tight stuffing box, causing undue friction, heating, and wearing of the piston rod.

• In the single-acting pump, because of the low pressure on the piston packing box, the leakage of ammonia past the piston rod is easily prevented. This in itself is an important saving.

Double-acting pumps on their downward stroke do not expel all the gas, owing to the fact that it is impracticable to use an arrangement of outlet and inlet valves to avoid large waste spaces. These clearance spaces are filled with compressed gas on the downward stroke; part of the charge that is compressed is not expelled, and expands back as the piston recedes, thus preventing the reception of a full charge on the next stroke. Another defect of a double-acting pump is that in case of accident the machine must be stopped until the part is repaired, a slight accident to one of the valves, pistons, or parts *disabling the whole machine*. In case of accident to our single-acting pump, the suction and discharge stop valves may be closed on the disabled pump only, and the other pump *may continue to run*, and in most cases, if speed is increased, *the work will go on the same as before*, the disconnected pump being examined or repaired at leisure.

The office of the ammonia condenser is to condense the ammonia gas as it is discharged from the compressors, and the work done in the condenser is the taking up of the latent heat of vaporisation, plus part of the heat generated by compression. The problem is to distribute the gas over such an area of absorbing or cooling surface that the latent heat is rapidly taken up by the cooling water and carried away. Insufficient condensing surface, or a bad distribution of gas, entails an increase of temperature inside the cooling pipes, and a corresponding high condensing pressure, thus taking more steam to drive the engine and a larger supply of condensing water, the loss from these sources being considerable. The arrangement of the condenser that secures the minimum temperatures within the coils with the least amount of condensing

Fig. 26.

water is the most efficient, preference being given to the simplest form.

Two styles are used by the Frick Company: the straight-pipe, open-air, or surface condenser shown in Fig. 26, or the round coils and circular tank submerged condenser, see Fig. 27. In both these condensers the coils are so arranged that an efficient distribution of the gas and lowest possible temperature with least amount of condensing water is secured. In both the hot gas is discharged in the upper portion of the coils and liquid drawn off as fast as formed from lower portion into the receiver, thus securing the full use and benefit of entire condenser - cooling surface. The choice of condenser depends upon the proposed location of same and quality of the condensing water. If the condenser is to be placed on the roof or in a room having weak floor timbers, because of its light weight, the open-air condenser is recommended, especially if the water is muddy and rapidly deposits a thick coating on the pipes, as the open-air condenser is easy of access and can be kept clean with a broom. If the water is clear and timbering strong enough to support a submerged tank in a closed room, or but little room to spare, it being more compact than the open style, the round-coil submerged condenser is used.

Although the ammonia - compression machine is so generally adopted, the design and construction of machines employing this system vary considerably. As an example of what one firm is doing we give below a description of a plant manufactured and designed by the Pulsometer Engineering Company, of 63 Queen Victoria Street, London, E.C. This firm has been in this business

7

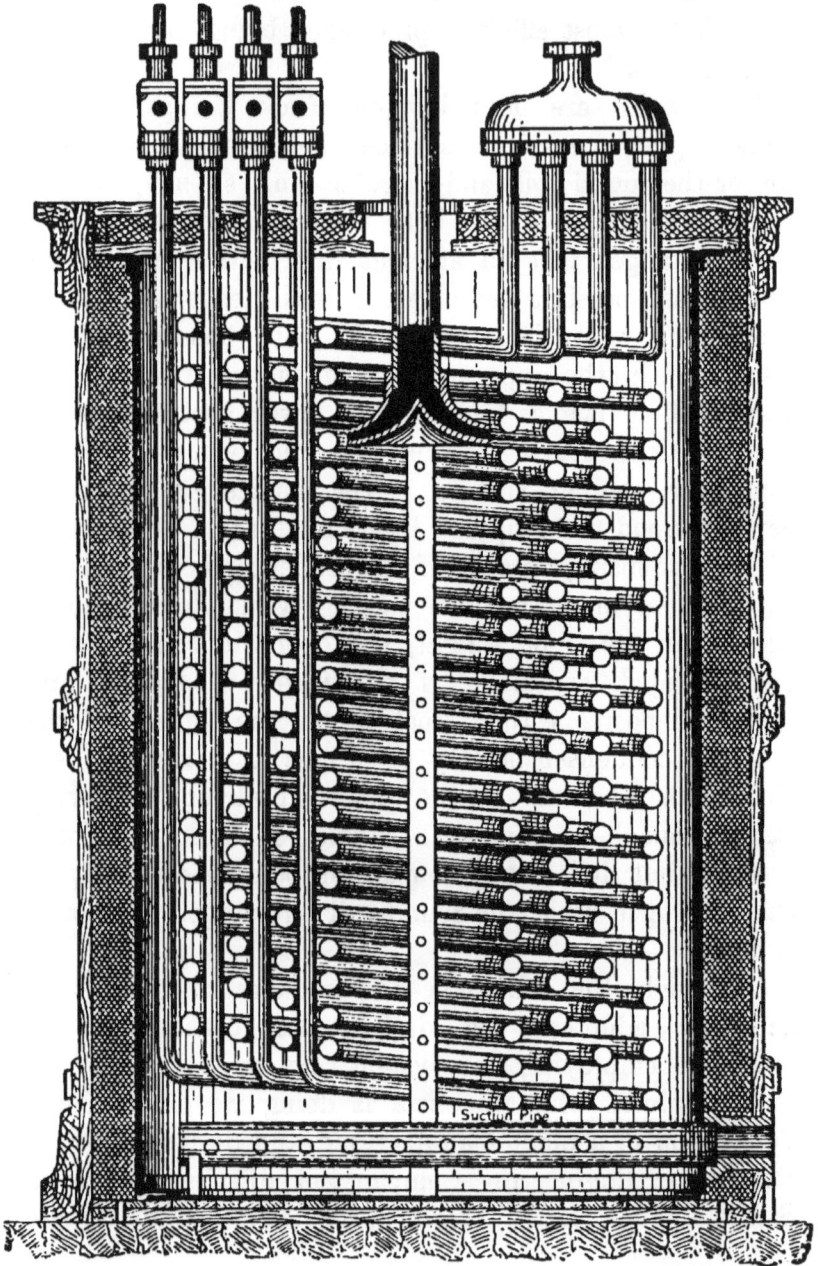

Fig. 27.

for some years, and has had great experience in nearly every branch of ice-making and refrigerating machinery, and has successfully carried out some of the largest installations in this country.

The leading features of these machines are—the highest possible economy in fuel and labour; extreme simplicity and ease of working; minimum loss of refrigerant; ample facilities for examination and repairs; the absence of all ammonia coils, except in the very smallest machines; no cast-iron in contact with the ammonia, except the compressing pump; small space required; very large and efficient surfaces in condenser and refrigerator; perfection of details.

This firm, with a view to further simplifying their machines, have recently patented and produced a new form of compressor, combined with a compound engine, complete on one bed, and which reduces the number of moving parts by about half.

Besides the method of making ice with distilled water in cans, this firm has another system in which the ice is formed direct in the freezing cells. Ice can be produced in blocks, 6 feet × 16 feet × 10 to 8 inches thick, and perfectly clear and transparent. With regard to the cooling of rooms, their usual practice is to use a system of special brine pipes placed in the cold room, and through these pipes cold brine is circulated. The advantage of this system is that the brine pipes hold a large quantity of cold brine, and thus the cooling is continued even after the machine is stopped. This storage of cold is advantageous in keeping the rooms at a much more uniform temperature than if the direct-expansion system be used.

The following is a description of one of their plants put down for producing a high-class ice at a low cost. The plant, though rated at 175 tons per week, repeatedly has turned out 177 tons; and it is no unusual thing for the plant to be run at full speed for three months day and night continuously without even stopping or slowing down.

When working to the maximum capacity the HP required for driving *everything*, including cranes, crushers, elevators, water and brine pumps, does not exceed 80, and this on a basis of 2 lbs. of coal per HP would give an efficiency of about 17 tons of ice made, not melted, per ton of coal.

Description of plant :—

The ice factory is divided into three rooms—first the boiler house, then the engine room, and finally the congealing room, the system being so arranged that the water, coal, stores, etc., come in at one end and the finished products go out at the other with the least amount of handling. The boiler room contains a large double-effect distiller, capable of dealing with 25 tons of water per day, for use when it is required to make clear ice. The boilers are two in number—one being used for the distiller and one for the engine—and are of the tubular type set in brickwork. The working pressure is 80 lbs. per square inch. There is also a spare main boiler.

A large tank is fitted under the boiler house floor holding 250 tons of water in reserve, and is connected with a small pump, capable of pumping the water to either of the boilers, distiller, and mould-filling tank.

The engine room, which opens out of the boiler house,

contains a large horizontal tandem engine, fitted with surface condenser, and drives the ammonia compressor by means of gear running in oil, the engine running twice the speed of the pump.

The ammonia pump is of the horizontal, double-acting type, cast in special metal, and tested with a high air pressure under water to insure that absolute tightness which is absolutely essential in this class of work. The inlet and outlet valves of the pump are of steel, working in steel cases, and without springs or buffers. The stuffing box round the compressor piston rod is provided with an inner and outer packing, leaving an annular chamber round the rod. This chamber is in communication with two pipes with a reservoir placed above the pump and filled with oil; any gas passing the back packing escapes into the annular space and thence into the top reservoir, where it is drawn off by the pump. By those simple means an easy and tight working rod is obtained.

The gas condensers are placed in a tank in the freezing room, and consist of special wrought-iron lap-welded U tubes connected with forged-steel T pieces, so arranged that any tube can be withdrawn. The advantages of this construction are great, for should a tube by any chance become defective it can easily be replaced, whereas if coils are used, in case of any defect, the whole coil must be removed and replaced, entailing an immense amount of expense and delay. The refrigerators are placed in tanks made of wood, and consist of welded-steel shells with forged-steel tube plates. The tubes are of wrought-iron, U-shaped, and secured into the tube plates with special metallic joints, and so arranged that they can easily be replaced or withdrawn..

The liquid ammonia surrounds these tubes, and the brine is circulated through the tubes by brine pumps placed in the same tanks. The power required to drive the pumps is extremely small, as the brine has only to be raised a few inches, when it will flow back to the congealing tank. The pumps are worked by gear from the main engine. It will be seen that the refrigerators are placed in separate tanks, and not in the ice tanks. They are thus better protected, and are much easier of access.

The auxiliary engine is placed in the corner of the engine room, and by a neat arrangement of straps, the line shaft, from which are driven the brine and circulating pumps, the elevator, crusher, and cranes, can be run either from the main or auxiliary, as desired, without stopping the plant.

The circulating pumps are of the vertical double-acting type, drawing the condensing water from a reservoir, and discharging into the gas-condenser tank, whence it flows to the engine condenser.

The two ice tanks are constructed of wood, to reduce the loss by radiation, etc., and contain moulds having a total capacity of 75 tons. These moulds are of galvanised steel, and produce blocks of about $1\frac{3}{4}$ cwt. Each tank is provided with a rope-driven travelling crane; the working levers are arranged for one man to work, and they can be stopped and started from anywhere down the centre of the house. The filling gear has been specially designed for quickly and easily filling the moulds, and is supplied with water from a large storage tank. The method of lifting and tipping the moulds is as follows :—The lifting frame is lowered by the crane on to the moulds in the freezing tank, and then by the movement of one lever takes hold of eight moulds at once, which are then lifted into the thawing-off bath, and thence into the tipping frame; the lifting frame is then removed, and, by the turning of a handle, the eight moulds are tipped and the ice discharged, the moulds being then replaced and refilled.

The whole of the connecting pipes between the various parts of the machine are, where in contact with ammonia, of solid steel tubing, and the valves and cocks are made from steel forgings.

It has been found, in a plant of this description, that the cost of coals, per ton of ice made, is not more than one shilling, and also that the entire labour, including getting the ice out, is only one shilling and three-pence per ton, the cost of the ammonia being under one farthing per ton of ice made.

In addition to those already described, ammonia-compression machines are manufactured in this country by Messrs. Siebe, Gorman, & Co., and by numerous other firms on the Continent and in the United States, but as they possess no special or distinctive features, it is unnecessary to enter into particulars regarding their construction.

System C—known as the absorption, was introduced about 1850. The principle employed is chemical or physical rather than mechanical, and depends on the fact that many vapours of low boiling point are readily absorbed by water, but can be separated again by the application of heat to the mixed liquid. A considerable number of machines in which ammonia was used in

combination with water as an absorbent were made by
Carré in France; but no very high degree of perfection
was arrived at, owing to the impossibility of getting an
anhydrous product of distillation; the ammonia distilled
over containing about 25 per cent. of water, which caused
a useless expenditure of heat during evaporation, and
rendered the working of the apparatus intermittent.

Ordinary ammonia liquor of commerce of ·88 specific
gravity, which contains about 38 per cent. by weight of
pure ammonia and 62 per cent. of water, is introduced
into a vessel named the generator.   This vessel is heated
by means of steam circulating through coils of iron
piping, and a mixed vapour of ammonia and water is
driven off.   This mixed vapour is then passed into a
second vessel, in order to be subjected to the cooling
action of water; and here, owing to the difference
between the boiling points of water and ammonia, fractional condensation takes place, the bulk of the water,
which condenses first, being caught and run back to the
generator, while the ammonia in a nearly anhydrous state
is condensed and collected in the lower part of the vessel.
The liquefying pressure for any given temperature is
shown in Fig 1.

This process of fractional condensation was introduced
by Rees Reece in 1867, and forms an important feature
in the modern absorption machine.   In the improved
form of apparatus, ammonia is obtained in a nearly
anhydrous condition, and in this state passes on to the
refrigerator.   In this vessel, which is in communication
with another vessel called the absorber, containing cold
water or very weak ammonia liquor, evaporation takes
place, owing to the readiness with which cold water or

weak liquor absorbs the ammonia, water at 59 deg.
Fahr. absorbing 727 times its volume of ammonia vapour.
The heat necessary to effect this vaporisation is abstracted
from brine or other liquid, which is circulated through
the refrigerator by means of a pump. Owing to the
absorption of ammonia, the weak liquor in the absorber
becomes strengthened, and it is then pumped back into
the generating vessel to be again dealt with as above
described. Though of necessity the various operations
have been described separately, the process is a continuous one, strong liquor from the absorber being constantly
pumped into the generator through the heater or economiser, while nearly anhydrous liquid ammonia is being
continually formed in the condenser, then evaporated in
the refrigerator and absorbed by the cool weak liquor
passing through the absorber.

Putting aside the effect of losses from radiation, etc.,
all the heat expended by the steam in the generator will
be taken up by the water passing through the condenser,
less that portion due to the condensation of the water
vapour in the analyser, and plus the amount due to the
difference between the temperature of the liquid as it
enters the generator and the temperature at which it
leaves the condenser. In the refrigerator, the liquid
ammonia in becoming vaporised will take up the precise
quantity of heat that was given off during its cooling and
liquefaction in the condenser, plus the amount due to the
difference in heat of vaporisation, owing to the lower
pressure at which the change of state takes place in the
refrigerator, and less the small amount due to the difference in temperature between the vapour entering the
condenser and that leaving the refrigerator, less also the

amount necessary to cool the liquid ammonia to the refrigerator temperature.   When the vapour enters into solution with the weak liquor in the absorber, the heat taken up in the refrigerator is imparted to the cooling water, subject also to corrections for differences of pressure and temperature.   Supposing there were no losses, therefore, the heat given up by the steam in the generator, plus that taken up by ammonia in the refrigerator, would be precisely equal to the amount taken off by the cooling water from the condenser, plus that taken off from the absorber.

This apparatus was afterwards improved by Stanley, who introduced steam coils for causing the evaporation in the generator; and then by Pontifex & Wood, who have succeeded in bringing the absorption machine to a considerable state of efficiency.

Their machine, as applied to the cooling of liquids, is shown in Fig. 28.   It consists of a number of strong cast-iron cylindrical vessels connected together by pipes and cocks.   The first of these is a large horizontal vessel called the generator, G, into which a charge of commercial liquor ammonia is placed.   This vessel contains a coil of steam pipes heated by steam from the ordinary steam boilers, so as to evaporate the ammonia which rises up a vertical cylinder called the separator, S, placed on the top of the generator.   This separator is so constructed that any watery vapour rising with the ammonia is condensed and returned to the generator.   From the top of the separator a pipe conveys the gas to the condenser, D, consisting of a number of coils of pipes, contained in a wrought-iron vertical cylinder, which is kept full of water in circulation. In this condenser the evaporated ammonia gas is condensed into a liquid form by the pressure caused by its own accumulation.   This liquid next passes to the cooler or refrigerator, R, which is a vertical cast-iron vessel, containing coils of wrought-iron pipes.   In the refrigerator the liquid ammonia, which leaves the condenser at a

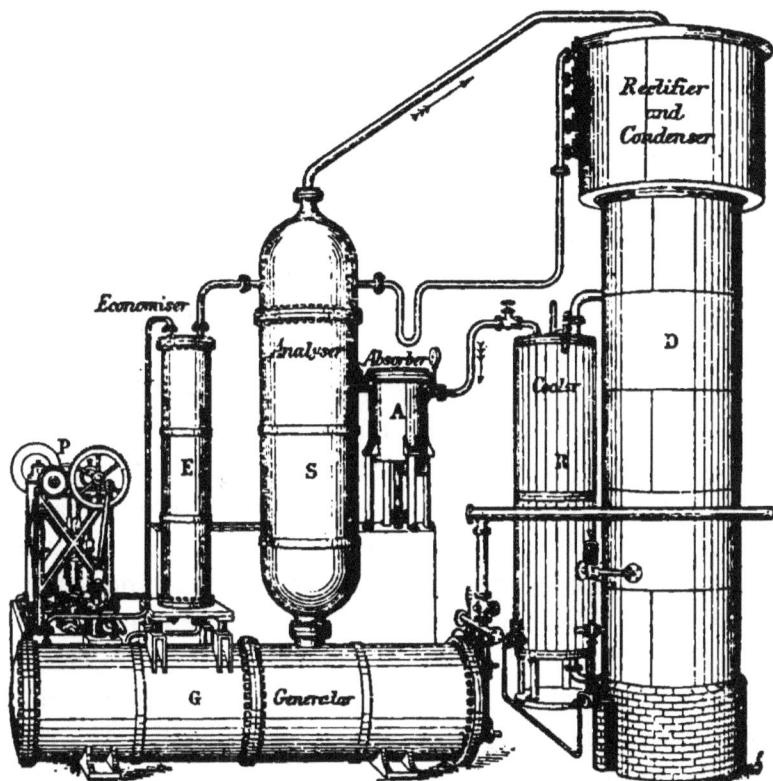

Fig. 28.

temperature of about 70 deg. or 80 deg. Fahr., is allowed to expand into gaseous form.   In doing so its sensible heat is rendered latent, and its temperature is ordinarily reduced down to about 10 deg. to 20 deg. Fahr., or, say, 22 deg. to 12 deg. of frost, but it can be reduced to a much lower point if desired. A circulation of water or brine is run through the coils of pipes in the refrigerator, and the expanding ammonia gas cools this water or brine down to any desired temperature. After doing this, the ammonia gas passes away through a pipe into another vertical cylinder called the absorber, A, in which it meets with and is absorbed by the water from which it was first evaporated in the generator. From the absorber the liquid ammonia is drawn by pumps, P, which force it back through an economiser or interchanger, E, into the generator, thereby raising its temperature by means of the water which is running from the generator into the absorber. From this generator it is evaporated again, and the operation is continuous, the same ammonia and water being used indefinitely.

The method of working is as follows :—After all the connections are made, the machine is started by filling the generator G with the ordinary ammoniacal liquor of commerce, and a little steam is admitted into the coil of pipes inside the generator, so as to raise just sufficient pressure of gas to expel all the air in the machine through a valve provided for the purpose on the absorber E.

When all the air is thus expelled, the full pressure of steam is turned into the generator coil. The ammonia in the solution being very volatile is immediately driven off in the form of gas, and passes through the separator B into the top of the condenser D, the water of the solution being left behind in the generator.

The condensing water is admitted at the bottom of the condenser and run off at the top, the ammoniacal gas passing down through a coil of pipe contained in the condenser.

The upper part of this coil is called the rectifier, and is fitted at intervals with traps or pockets. The gas, passing down the coil, is cooled by the condensing water, and parts with any watery particles that may have been carried over with the ammonia. This water is caught in the traps and is at once passed out of the coil and returns into the separator S.

The ammoniacal gas, after passing the lowest trap, is quite dry or anhydrous, and it continues to pass on down the coil in the condenser until by its accumulation it reaches a pressure at which the gas becomes liquefiable, the liquefaction being greatly assisted by the reduction of temperature due to the condensing water used.

We have now obtained liquid anhydrous ammonia, and the apparatus is so arranged that, as the gas becomes liquefied, it passes into the refrigerator or cooler R.

The ammonia in this vessel, being quite free from water, vaporises, under the ordinary pressure of the atmosphere, at a temperature as low as 60 deg. below freezing point. At any higher temperature it passes freely into a gaseous form, and, at the moment of thus changing its form, it absorbs and renders latent an immense amount of heat.

The only source from whence it can abstract this heat is from the contents of a coil of pipe provided for that purpose in the cooler cylinder.

In breweries, the water to be cooled is passed direct through this coil ; and for ice-making, a strong solution of chloride of calcium, or brine, as it is called, is passed through it, which, after being cooled to a very low temperature, is pumped to the ice boxes, there to freeze the water and convert it into ice, returning again to the machine to be recooled for further use.

The ammonia, now again in the gaseous form, passes from the top of the cooler into the absorber A. A pipe connects this vessel with the bottom of the generator, through which the pressure in the latter forces a constant stream of the water left in it at starting into the absorber. This water, containing little or no ammonia, greedily absorbs the gas coming from the cooler, and the two form a strong solution of ammoniacal liquor similar to that originally put into the generator. This solution is then drawn away by one of the pumps P, and forced through a coil of pipe in the economiser or heater E into the top of the separator. The solution, now rich in ammonia, then passes down the cylinder through a series of trays; these trays being heated by the hot vapour rising from the generator, the ammonia is again separated from the water in which it is dissolved, and the solution gradually becomes weaker until it falls into the generator almost entirely exhausted of ammonia.

The ammonia, now once more in the form of gas, passes into the condenser as before, to be again made dry and liquefied, thence to the cooler, where by its reconversion into vapour it again produces the cold, and passes once more back to the absorber and pumps.

Thus the whole process forms a continuous cycle, the same changes from liquid to gas and gas to liquid being constantly repeated, with no destruction of material whatever except the quantity of coal consumed under the boiler.

The economiser or heater E utilises the heat of the water as it passes from the generator on its way to the absorber by heating up the ammoniacal solution before it enters the separator, and so saves steam ; whilst at the same time the reduced temperature of the water enables it to reabsorb a larger proportion of ammoniacal gas.

The sources of loss in such an apparatus are :—

Radiation and conduction of heat from all vessels and pipes above normal temperature, which can, to a large extent, be prevented by lagging.

Conduction of heat from without into all vessels and pipes that are below normal temperature, which can also, to a large extent, be prevented by lagging.

Inefficiency of economiser, by reason of which heat obtained by the expenditure of steam in the generator is passed on to the absorber, and there uselessly imparted to the cooling water.

The entrance of water into the refrigerator, due to the liquid being not perfectly anhydrous.

The useless evaporation of water in the generator.

With regard to the amount of heat used, it will be evident that the whole of that required to vaporise the ammonia, and whatever water vapour passes off from the generator, has to be supplied from without. Owing to the fact that the heating takes place by means of coils, the steam passed through may be condensed, and thus each lb. can be made to give up some 950 units of heat. With the absorption process, worked by an efficient boiler, it may be taken that 200,000 thermal units per hour may be eliminated by the consumption of about 100 lbs. of coal per hour, with a brine temperature in the refrigerator of about 20 deg. Fahr.

It will have been seen that the heat demanded from the steam is very much greater in the absorption system

than in the compression.   This is chiefly due to the fact
that in the absorption system the heat of vaporisation
acquired in the refrigerator is rejected in the absorber:
so that the whole heat of vaporisation required to
produce the ammonia vapour prior to condensation has
to be supplied by the steam.   In the compression system
the vapour passes direct from the refrigerator to the
pump, and power has to be expended merely in raising
the pressure and temperature to a sufficient degree for
enabling liquefaction to occur at ordinary temperatures.
On the other hand, a great advantage is gained in the
absorption machine by using the direct heat of the steam
without first converting it into mechanical work ; for in
this way its latent heat of vaporisation can be utilised
by condensing the steam in the coils and letting it escape
in the form of water.   Each lb. of steam passed through
can thus, as has been shown, be made to give up
some 950 units of heat ; while in a steam engine, using
2 lbs. of coal per indicated horse power per hour, about
160 units only are utilised per lb. of steam, without
allowance for mechanical inefficiency.   In the absorption
machine also the cooling water has to take up about
twice as much heat as in the compression system, owing
to the ammonia being twice liquefied, namely, once in the
condenser and once in the absorber.   It is usual to pass
the cooling water first through the condenser and then
through the absorber.

The cost of producing clear block ice in this country,
with an absorption machine of 15 tons capacity per
twenty-four hours may be taken at about 4s. per ton, with
good coals at 15s. per ton, exclusive of allowance for
repairs and depreciation.   About 10 tons of ice can be

made per ton of coal consumed, assuming an evaporative duty of 8 lbs. of water per lb. of coal.

System D. In this, known as the binary absorption, liquefaction of the refrigerating agent is brought about partly by compression and partly by absorption; or else the refrigerating agent itself is a compound of two liquids, one of which liquefies at a comparatively low pressure, and then takes the other into solution by absorption.

An apparatus of the first kind was brought out in 1869 in Sydney by Messrs. Most & Nicol, who used ammonia, with water as an absorbent. The machine consisted of an evaporator or refrigerator, a pump, and an absorber. The evaporator was supplied with strong ammonia liquor, which was vaporised by means of the reduction of pressure induced by the pump, and so abstracted heat from the liquid to be cooled. The weak liquor passing out at the bottom of the evaporator was led by pipes to the pump, where it met with the ammonia vapour, along with which it was forced through cooling vessels under sufficient pressure to cause the solution of the ammonia : and the strong liquor thus formed was again passed into the evaporator. This machine was only used by the inventors in Australia, so far as the author is aware ; and he has no particulars as to fuel consumption or cost of working. It was not likely, however, to be a very economical apparatus, because the whole of the water entering the evaporator with the ammonia had to be reduced in temperature, giving up its heat to the ammonia vapour, and to that extent preventing the performance of useful cooling work. But this disadvantage was in some degree compensated for by reducing the temperature of the strong liquor

8

before it entered the evaporator, by means of an interchanger, through which the very cold weak liquor passed
on its way to the pump.

In machines of the second kind, in which both liquids
are evaporated at a low temperature, the foregoing objection does not exist; and, though this mode of working
has not as yet been introduced into this country, it has
been successfully employed in the United States for
several years by Messrs. D. Motay & Rossi.   The liquid
used is a mixture of ordinary ether and sulphur dioxide,
and has been termed ethylo-sulphurous dioxide; its
adoption being decided on after a series of experiments
with numerous other combinations of ethers and alcohols
with acids.   In these investigations it was found that
liquid ether at ordinary temperatures possessed an
absorbing power for sulphur dioxide, amounting to some
300 times its own volume; while at 60 deg. Fahr. the
tension of the vapour given off from the binary liquid
was below that of the atmosphere.   In working, both liquids
evaporate in the refrigerator, under the influence of the
pump; and in the condenser the pressure never exceeds that
necessary to liquefy the ether.   The compressing pump has
less capacity than would be required for ether alone, but
more than for pure sulphur dioxide.   The author has no
particulars as to the cost of making ice by this process;
but he believes it to be somewhat less than with ether.

An interesting application of the binary system has
lately been made by Raoul Pictet, who found that by
combining carbon dioxide and sulphur dioxide he could
obtain a liquid whose vapour tensions were not only
very much less than those of carbon dioxide (see Fig. 1),
but were actually below those of pure sulphur dioxide at

temperatures above 78 deg. Fahr. This is a most remarkable and unlooked-for result, and may open up the way for a much greater economy in ice production than has yet been attained. As to the results that have been obtained with this process, the author has no definite particulars; but he understands it is stated to give a production of 35 tons of ice per ton of coal.

The following systems are usually adopted for transferring the cold generated by the refrigerating machinery to the chambers or rooms requiring to be cooled.

An uncongealable solution of salt or chloride of lime in water is reduced by the refrigerating machine to a low temperature, and this liquor acts as transmitter of cold in one or other of the undermentioned methods :—

1. The cold brine is constantly circulated from the brine refrigerator through pipes placed in the cold chambers, and returned to the brine cooler, the result being that not only is heat abstracted from the air of the refrigerated rooms, but also a large degree of the moisture which may be present in them, this moisture being condensed on the exterior of the brine-pipe systems either in the form of condensed water or hoar-frost. Suitable drip trays are provided, if required, to prevent this moisture from falling upon the contents of the rooms. The circulation of air with this system is a moderate one, being produced merely by the differences between the temperatures prevailing near the brine pipes and those in the lower parts of the rooms.

2. The brine is cooled in a shallow rectangular open tank containing the evaporator coils. On the tank is mounted a number of slowly-revolving transverse shafts, and on each shaft is fixed a number of parallel discs, partly

immersed in the brine, the entire apparatus being placed in an insulated passage through which an air current is continually passed by a fan, in a direction parallel to the revolving discs.   It will be seen that as the discs revolve and are kept covered by a film of the refrigerated brine, the air passing between the disc spaces becomes cooled, and produces a low temperature in any chamber or room into which it may be conducted through properly arranged air trunks.   As a rule, the air is always taken back from the cold rooms, and any required amount of fresh air is introduced by means of adjustable openings in the air trunks, communicating with the outer atmosphere.   In this instance, also, moisture may be removed from the refrigerated rooms and deposited in the brine contained in the trough.   No accumulation of frost can take place, and the refrigerated surfaces are always perfectly active.   The circumstance of all moisture being deposited in the brine necessitates either a periodical loss of the same or its reconcentration.   The fan produces a very effective air circulation within the rooms to be cooled. This, in most cases, is extremely desirable, and, as will be readily understood, produces the most beneficial results.

3. Instead of using an uncongealable liquid as bearer of the cold, the refrigerator coils, in which the vaporisation of the anhydrous ammonia takes place, are sometimes constructed with extra large surfaces, and placed either in the upper part of the rooms to be cooled, or in a separate chamber.   In the latter case, a fan constantly circulates the air between this chamber and the refrigerated rooms. This is the system generally adopted on board ships, and it has been found to be in all respects most satisfactory.

In cases where the air temperature is not sufficiently high to cause a complete removal of the snow deposited on the ammonia coils, the snow is thawed by the ammonia vapours themselves, the evaporator coils being for the time used as a condenser. Occasionally the snow is thawed by a current of hot air taken from the outside.

Though all the foregoing systems and apparatus have been applied on an extensive scale in actual practice for the refrigeration of storage and freezing rooms, the system most strongly recommended, in cases where the application is possible, is the combination of revolving discs immersed in brine. It is a very simple and compact arrangement, involving no expensive repairs. There are no brine or ammonia pipes in the rooms; whilst the rapid air circulation by the fan is easily managed, and has been found in most cases to be requisite for obtaining a satisfactory result as to purity, dryness, and equable temperature in all the rooms.

Where circumstances require the refrigerated rooms to be at a distance from the refrigerating machine, it is generally most convenient to place bundles of brine pipes into each room; but even in such a case, in the event of a small amount of motive power being available close to such rooms, the system of revolving discs and fans can be readily applied, the brine being cooled in a refrigerator near the compressor, and conveyed to and from the disc tanks through insulated pipes.

CHAPTER IV

Compression and Expansion of Air—Energy of a Gas—Actual, Adiabatic and Isothermal Curves for Air—Rate of Expansion of Air—Efficiency of Machine—Moisture in Air—Theoretical Diagrams modified in Actual Practice—Description of Cold-Air Machine—Kirk—Giffard—Bell-Coleman—Sturgeon—Hick Hargreaves & Co.—J. & E. Hall—Haslam—and Lightfoot—Cold-Air Machines.

THE intrinsic energy of a permanent gas, or its capacity for performing work, depends entirely upon its temperature. Increase of pressure imparts no additional energy, but merely places the gas in such a condition relatively to some other pressure as to enable advantage to be taken of its intrinsic energy by expansion. Thus a pound of air at ordinary atmospheric pressure has the same intrinsic energy as a pound of air at 50 lbs. pressure above the atmosphere, so long as their temperatures are the same ; but in the former case no part of the energy can be made use of by expansion without the removal of at least a part of the equal and opposite resistance of the atmosphere, while in the latter case expansion can take place freely until the pressure is reduced to that of the atmosphere. As mechanical work and heat are mutually convertible, it is obvious that, if during expansion a gas is caused to perform work on a piston, its supply of heat must be drawn on to an extent measured by the thermal equivalent of the work done, provided no extraneous source of heat exists from which the deficiency can be made good; and the gas after expansion will be colder than it was before expansion. Expansion behind a piston, without the addition of heat from any extraneous source, is called adiabatic expansion; and the following are the relations between temperature, volume, and pressure for any two points in the same adiabatic curve:—

$$\frac{t}{t_1} = \left(\frac{V_1}{V}\right)^{r-1} = \left(\frac{P}{P_1}\right)^{\frac{r-1}{r}}$$

where t and V and P denote absolute temperature, volume and absolute pressure before expansion, and t_1 V_1 P_1 those after expansion, while r is the ratio of the specific heat under constant pressure to that with constant volume. In the case of atmospheric air r=1·408. Absolute zero = − 461 F. The results with regard to the expansion are shown graphically by the curve E F in Fig. 29, page 121 : E D representing the initial volume at a pressure of 50 lbs. per square inch above the atmosphere and at a temperature of 70 deg. Fahr., which is expanded down to the volume F B at atmospheric pressure, the temperature falling to − 115 deg. Fahr. owing to the performance of the mechanical work represented by the area E F I K.

During adiabatic compression, the converse results take place, and the same relations exist between absolute temperature, volume, and absolute pressure as during expansion: t_1 V_1 P_1 denoting those before compressing, and t V P those after compression. On the diagram this is shown graphically by the curve A C, which is the

adiabatic for the compression of the initial volume A B
at atmospheric pressure and 70 deg. Fahr., to 50 lbs. per
square inch above the atmosphere, the temperature rising
to 354 deg. Fahr. owing to the acquirement of heat due
to the performance of the mechanical work represented
by the area A G J C. A E is the line of isothermal
compression and expansion between the two points
A and E.

In the succeeding remarks reference will be made to
the use of ordinary atmospheric air alone ; for, although
in one or two special instances this class of machinery
has been applied to the cooling of some of the more
volatile hydrocarbons, its almost universal application at
the present time is for the cooling of air, which, therefore,
will alone be dealt with.

When atmospheric air is compressed under a piston,
without either loss or gain of heat from without, it is
raised in temperature, mechanical work expended on the
piston being transferred to the air in the form of heat.
If this compressed and heated air, at that pressure and
temperature, be then introduced below another piston, and
expanded without loss or gain of heat from without down
to its original pressure, it will also resume its original
temperature, and will have given back, while expanding,
useful work precisely equal in amount to that absorbed
during compression.   If, however, after compression, the
air is first cooled, by allowing some of its sensible heat to
be absorbed by some cooler substance, and is then expanded under a piston to atmospheric pressure, a less
amount of useful work will be given back than in the
first case, and the air, after expansion, will be found to
occupy less than its original volume, and to be colder

than its original temperature by a difference which is greater or less, according as the quantity of heat taken away before expansion is large or small.

The operation just described forms the basis upon which cold-air machines are constructed. In its simplest form it is shown graphically in Fig. 31, where A B represents a volume of atmospheric air, considered for the sake of convenience, as a perfect gas, at a temperature of

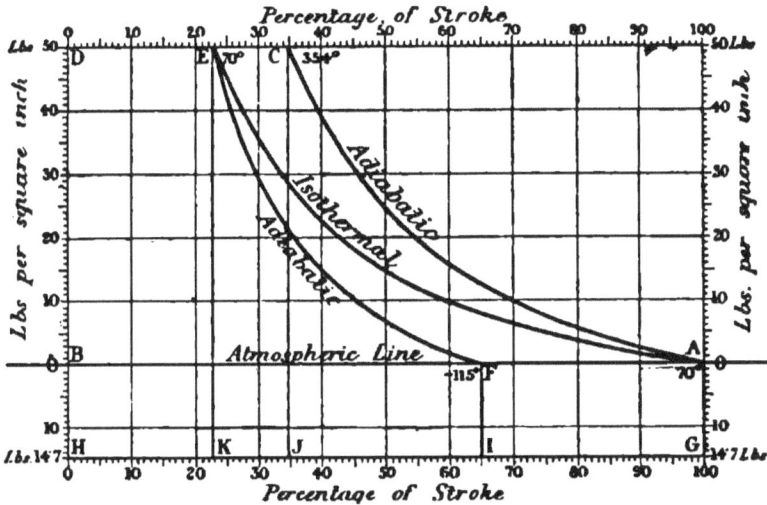

Fig. 29.

52 deg. Fahr. This air is compressed under a piston to the volume C D, according to the adiabatic curve B D. The pressure A C above the atmospheric pressure is then, in the present example, 50 lbs. per square inch, or 65 lbs. absolute; and the temperature is 321 deg. Fahr., giving a rise of 269 deg. Fahr. Now suppose that, instead of immediately expanding the volume C D of hot compressed air back to atmospheric pressure, we first abstract a

portion of its sensible heat, and so reduce its temperature
to 52 deg. Fahr.; it will then be found that its volume is
also reduced to C E, where C E bears the same ratio to
C D as the new absolute temperature bears to the old;
or, taking —461 deg. Fahr. as the absolute zero of
temperature:—

$$C E : C D :: 513 : 782.$$

On now expanding the volume C E to its original
atmospheric pressure, the piston will be pushed out only
to the position G, and the final temperature of the air
will be 125 deg. below zero Fahr. The efficiency of the
operation is represented by

$$\frac{\text{Volume swept through by expansion piston}}{\text{Volume swept through by compression piston}}$$

and the area B D E G gives the theoretical mechanical
force required for driving the machine.

This force will of course be greater as the extent of
compression is greater; but, on the other hand, assuming
the temperature of the cooling agent, which is generally
water, to be constant, the cold produced by expansion will
be correspondingly greater.

Commencing then with the simple fact that air is
heated by compression and is cooled to a like amount by
expansion, it next becomes of importance to ascertain how
far the presence of water, in the condition either of steam
or of mist or of actual liquid, affects the heating or cooling
of air, and the conditions of working any given apparatus:
in addition to its effect in the formation of ice, which is
very objectionable.

The important fact to be noted in this connection is
that air at constant pressure, having free access to water,
will hold a different quantity of water in solution as

vapour or steam, at each different temperature ; or con-
versely, the temperature of the " dew point " for any body
of air varies with the quantity of water held in solution
by it. The hotter the air, the more water can it hold
without depositing. See Fig. 30.

Thus if air is highly heated, and water is then admitted
to it in the form of spray or injection, it will take up
much more water before becoming saturated than it could
have held before it was thus heated. Again, if air, under

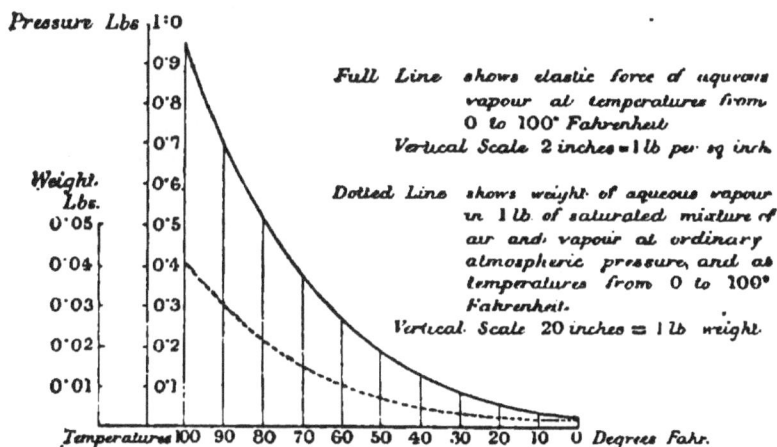

Fig. 30.

compression and saturated with vapour, is allowed to
expand, a large quantity of its contained vapour will condense and freeze into snow, thereby yielding up a quantity
of heat to the air, which air is in consequence cooled less
in expanding than it would have been had it been dry
air to start with. This · freezing is also a practical evil,
from the deposition of ice about the valves and in the air
passages, which necessitates frequent stoppages even in
small machines. An appreciation of these facts will

render it easy to understand the action of the various machines about to be referred to, in which much depends upon the presence of water in the air at different times.

The amount of aqueous vapour present in the atmosphere varies from that required to produce saturation down to about one-fifth of that quantity. At any given temperature a volume of saturated air can contain only one definite amount of vapour in solution—see Table D; and, if from any cause additional moisture be present, it cannot exist as vapour, but appears as water in the form of fog or mist. Various means have been devised for ridding the air more or less completely of its contained moisture, in order to obviate as much as possible the practical evils resulting from its condensation and freezing, this being at one time considered one of the most important points in the construction of cold-air machinery. Experience, however, has since demonstrated that these evils were much exaggerated, and that the condensation of the vapour and deposition of the moisture in the ordinary cooling process after compression, which is common to every cold-air machine, are amply sufficient to prevent any serious deposition of ice about the valves and in the air passages : provided, firstly, that these valves and passages are well proportioned, and, secondly, that proper means are adopted for obtaining in the coolers a deposition of the condensed vapour, which would otherwise pass with the air into the expansion cylinder in the form of fog, and become converted into ice. If the compressed air be thoroughly deprived of its mechanically-suspended moisture, the amount of vapour entering the expansion cylinder is extremely small. Another matter from which the mystery has now been dispelled is the meaning of the term "dry" air, so much used by the makers of cold-air machinery. No doubt it is still to a large extent popularly supposed that, unless the air be subjected in the machine to some special drying process, it will be delivered from the expansion cylinder in a moist or damp state, and in consequence be unfitted for use in the preservation of perishable food, and for other purposes. But no such state could really exist; for, whether the air be specially dried or not, its humidity when delivered from the expansion cylinder is precisely the same, so long as its temperature and pressure remain the same, inasmuch as in practice it is always in a saturated condition for that pressure and temperature. The difference lies in the amount of ice formed, which of course is greater if the amount of moisture entering the expansion cylinder is greater; but this quantity, as has been already stated, may, in Mr. T. B. Lightfoot's opinion, be brought down within perfectly convenient limits by a proper construction of the cooling vessels.

In his latest machines, therefore, all special drying apparatus has been dispensed with; the air being simply compressed, passed through a surface cooler, and expanded back to atmospheric pressure. On the other hand, Messrs. Haslam & Co. of Derby, we understand, still apply an interchanger; while Messrs. J. & E. Hall of Dartford fit instead a centrifugal moisture separator, which gives good results.

Before passing on to the machines in which the cooling takes place by mechanical means, it will be well to show briefly how far the theoretical diagrams are modified in actual practice. In addition to the adiabatic curve B D, Fig. 31, is shown the isothermal curve B E, for the compression of a perfect gas, from atmospheric pressure

and 52 deg. Fahr. temperature, up to 65 lbs. per square
inch absolute pressure.   The former of these is the curve
which would be produced if the air could be compressed
instantaneously, or without transmission of heat; and the
latter the curve which would be produced if it could be
compressed without raising its temperature at all.   The
curves obtained in practice of course fall between the
two; the nearer they approach the isothermal line the

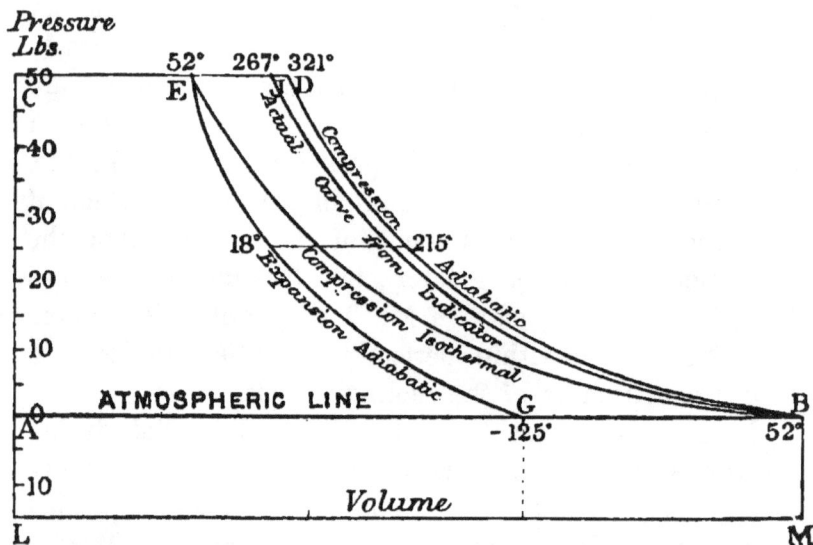

Fig. 31.

better.   The full line B J is a copy of the actual compression curve, in a diagram taken with a Richards indicator from the compression cylinder of a machine made
by the Linde British Refrigeration Co.   The initial temperature of the air entering the cylinder was 52 deg.
Fahr.; and it contained, as ascertained with a hygro-
meter, 0·007 lb. of aqueous vapour to the pound of
mixture, this being about 88 per cent. of saturation for

the observed temperature. By calculation from the volume, the temperature at the end of the stroke was 267 deg. Fahr. ; whereas, if the compression had been accomplished adiabatically, it would have been 321 deg. Fahr.

The air thus compressed is delivered to the cooling apparatus, consisting in this case of an arrangement of small brass tubes, having cold water flowing through them. The air passing round the outside of the tubes is thus reduced in temperature to within from 5 deg. to

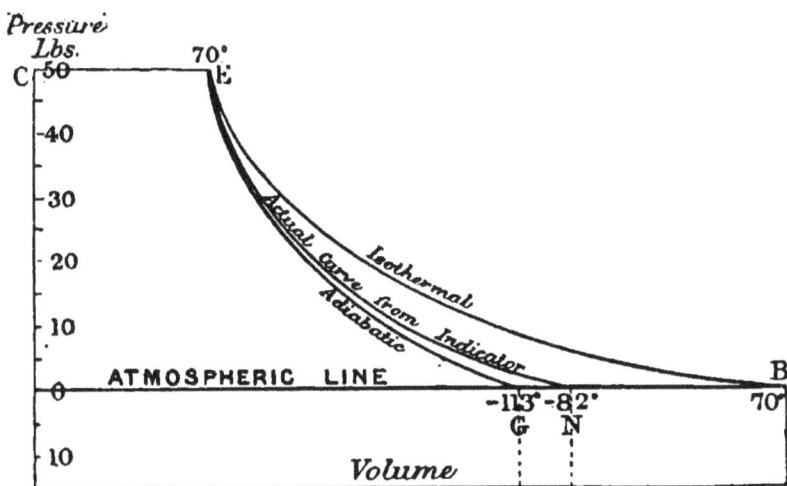

Fig. 32.

10 deg. of the initial temperature of the cooling water ; and with this abstraction of heat, its capacity to retain vapour being lessened, a portion of the moisture it contains is condensed, and may be collected and run off if suitable means be provided. In practice, with the machinery under the conditions mentioned above, the air, if cooled to 70 deg. Fahr., may be made to part with about one-half of its contained moisture at this stage.

In Fig. 32 are shown the adiabatic and isothermal lines of expansion, E G and E B respectively. The volume C E is the same as the volume C D in Fig. 20, corrected for reduction of temperature and for deposition of vapour. The intermediate full line E N shows as before the actual curve of expansion in an indicator diagram taken from the above machine. This curve, as might be expected, never falls to the adiabatic line, owing to gain of heat from without, and to the heat given off in the condensation and freezing of the moisture. In this case the final temperature was calculated at 82 deg. below zero Fahr.; whereas the temperature of the air expanded adiabatically would be 113 deg. below zero Fahr.

Fig. 33 is the plan of a horizontal cold-air machine suitable for marine purposes. The air is taken through the air inlet into the double-acting compression cylinder C, with gun-metal liner forming the water-jacket; this material being employed in preference to cast iron on account of its greater conductivity. This cylinder discharges the air, after being compressed to about 65 lbs. per square inch absolute, into the series of coolers B B B, containing rows of brass tubes, through the inside of which the cooling water is caused to circulate. Thence it passes to the expansion cylinder E, fitted with a trunk piston. Each end of this cylinder is fitted with distinct adjustable cut-off valves. After the air is expanded, it passes from the expansion cylinder through the air outlet pipe into the cooling chamber. A jacketed steam cylinder A, with adjustable cut-off, supplies the necessary driving power. The foregoing may be taken as a general description of all cold-air machines, the various makers differing from it only in matters of detail and construction.

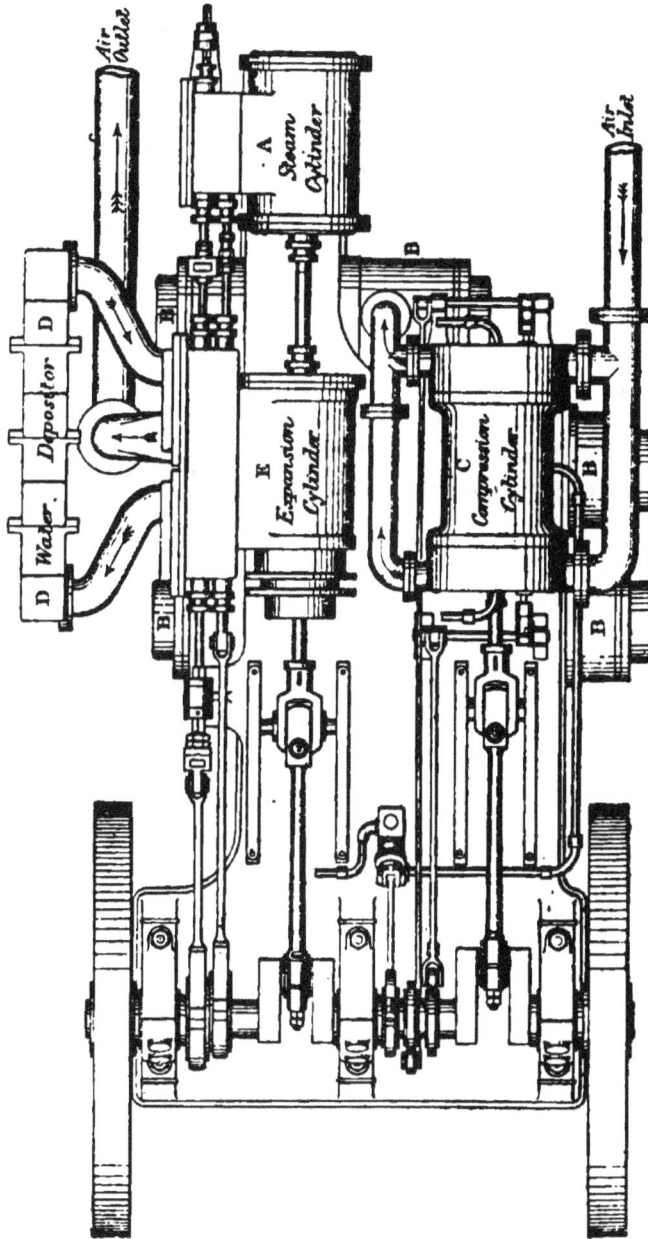

Fig. 33.

The following particulars of some cold-air machines, which have been brought more or less prominently into notice, may prove interesting and instructive. Dr. Kirk designed two kinds of cold-air machines, namely, dry-air and moist-air machines. The dry-air machine is simply Stirling's air engine reversed. It consists of a compression pump, a hot chamber, a displacer piston and regenerator, and a cold chamber. The hot and cold chambers form the two ends of a cylinder in which the displacer piston works. The operation of the machine is as follows:—By means of the pump, air is forced into the hot chamber, the heat formed by the compression being abstracted by metal kept cool by circulating water; when cool, it is moved by the displacer piston into the end of the cylinder forming the cold chamber. The compression piston now moves outward, expanding the air in the cold chamber, and producing an extremely low temperature. The liquids to be cooled or frozen are either placed in or passed through the cold chamber. As usually made, the compression cylinder is horizontal and double-acting, one cooling cylinder being placed at each end, and off the centre line of compression cylinder, so that the rods for working the displacer pistons pass clear of the bore. The covers of the cooling cylinder, as it is called, although it is really both a hot and cold cylinder, are hollow; in the bottom one, or hot end of the cylinder, cold water circulates, and in the top cover a saturated solution of salt in water. The covers are made corrugated on their inner surface, so as to have a large amount of surface. They are also made in two parts, the corrugated plate being preferably of brass, this substance being a good conductor of heat. The air in these machines must be perfectly dry, because if moisture were present it would freeze and choke up the regenerator. To prevent this, the air is passed over chloride of lime, a very deliquescent salt. In a machine of this kind, to make 3½ tons of ice in twenty-four hours the compressor is 18 inches diameter, with a stroke of 3 feet. The machine is driven at from 60 to 70 revolutions per minute. In order to reduce the size of the machine as much as possible, the volume of air enclosed in the compression and cooling cylinders is kept at 150 lbs. pressure. The power required to drive this machine is from 50 to 60 horse power, according to the temperature at which it is working. The difference in temperature between the brine going into the machine and that going out likewise varies, the higher the temperature at which the brine enters the machine the greater the difference, and also the greater is the cooling effect in proportion to the power used. For example, brine ingoing 19 deg. Fahr., outgoing 15 deg.; ingoing 15 deg. Fahr., out 11½ deg. These figures are taken from a machine working in a hot climate. The first machines of this kind sent abroad were driven by the ordinary non-condensing engine, and used about an equal weight of coal to the ice produced. By using brass corrugated plates and compound engines the consumption of coal is reduced to 10 cwts. per ton of ice.

Dr. Kirk's moist-air machine is practically a " Rider " air engine reversed, and consists of two cylinders placed side by side; one being the hot and the other the cold cylinder. The cylinders are connected at each end by a passage, in which are placed the regenerators, formed of several thicknesses of wire gauze. Through these both the hot compressed air and the cold expanded air pass, on their way from the one cylinder to the other ; so that there is a continual alternate compression and expansion of the air, and a continual heating and cooling of the regenerators. The high and low-pressure steam cylinders are placed above the air cylinders, and the piston rods of the former connect with and drive the plungers of the latter ; a grasshopper beam and connecting rod joins each cylinder with the crank shafts ; the cranks being set at an angle of 120 deg. The heat caused by compression is partially carried off through the cylinder cover, which is water-jacketed, and the cold from expansion is used to abstract heat from a current of brine or other medium, circulating over the cover at the expansion end.

We understand that these moist-air machines produce 1 ton of ice upon a consumption of 5 cwts. of coal ; which indicates a very high efficiency.

This machine, however, can scarcely be classed as a machine for producing cold air, since the cooled air is used only for abstracting the heat from some medium not in direct contact with it, and cannot be itself discharged for use. For this reason, the machine itself is more economical than those in which the cold air is directly made use of ; for the air, being used over and over again, assists in keeping down the temperature of compression, and thus reduces the amount of mechanical work required.

The Giffard cold-air machine consists of one single-acting, water-jacketed compression cylinder, and one single-acting expansion cylinder ; both are vertical and worked from cranks on an overhead shaft. The compressed air is led from the cylinder bottom into the cooler, which is merely a cluster of small tubes placed vertically in a case. The cooling water passes upwards outside the tubes, and thence goes to the compression cylinder jacket; the air is admitted into a casing below the ends of the tubes, passes up through them, and is taken off from the top to a wrought-iron reservoir. A pipe from this reservoir supplies the air to the expansion cylinder; the admission and exhaust being controlled by two independent steel mitre valves in the cylinder bottom, worked by cams from the shaft.

In this machine no attempt is made at drying the air; all the moisture taken into the compression cylinder is discharged in the form of snow from the expansion cylinder, with the exception of the portion deposited in the cooler owing to the partial cooling of the compressed air.

M. Giffard uses a special form of piston packing, made of two layers of indiarubber, the outer one hard and the inner soft. This packing, which is altogether about $\frac{3}{8}$-inch square in section, is inserted in a groove turned in the piston; and small holes, drilled from this groove to the underside of the piston, admit the air pressure to the back of the ring, thus making a similar joint to the ordinary cupped leather.

The Bell-Coleman refrigerator consists of an ordinary machine for producing cold air by compression, cooling and expansion, combined with an apparatus for depositing a portion of the moisture before the air is admitted to the expansion cylinder. In this system, the air is partially cooled during compression by the actual injection of cooling water into the compressor, and by causing the current of compressed air flowing from the pumps to come in contact with a spray of water. From the pumps, the mixed air and water is led by pipes into a chamber or chambers with perforated diaphragms, which catch a portion of the suspended moisture. The air, still in its compressed state, and cooled to within 5 or 10 degrees of the initial temperature of the cooling water, is then led to the expansion cylinder through the interior of a range of pipes, or other apparatus, with extended metallic surfaces, cooled externally to a lower temperature than that of the cooling water, so as to induce a further reduction in temperature and consequent deposition of moisture. This extra cooling of the compressed air is effected either by allowing the cold expanded air, before it reaches the chamber to be cooled, to come in contact with the outside of the range of pipes, or by exposing these pipes to the spent air passing from the cold chamber.

This system is objectionable at sea, from the corroding action of the salt water upon the cylinder, pistons, valves, etc., particularly when, as must often be the case, the machine lies idle for several days. Again, with internal injection there is a decided loss of efficiency, wherever it is possible to use the same air over and over again; which can be done by making the cold chamber into a comparatively air-tight compartment, and drawing from it the supply to the compression cylinder. With an external system of cooling the compressed air, full advantage is gained by this arrangement; for the expanded air discharged into the cold chamber, even if it be not passed through a dry-air machine, becomes practically free from moisture, when discharged in the usual way through some kind of trap for collecting the particles of snow. This air, being continually compressed and expanded, free from all contact with water, ensures economy in working, firstly, by utilising what may be termed the waste cold, and, secondly, by excluding water vapour, which otherwise would have to be condensed and deposited, with a consequent loss of power. Thus, after a few cycles of· operation, the whole of the moisture is removed from the air, which works thenceforward like a perfect gas. With internal injection, on the contrary, though the waste cold can be utilised, there still remains a continual loss from the saturated condition in which the air, even if used over and over again, must necessarily be delivered from the cooling apparatus on each occasion.

The difficulties in working this machine as a dry-air refrigerator may be further seen by considering its performance in a tropical climate, where, even at sea, the water available for cooling the compressed air would probably have an initial temperature of 90 deg. Fahr. Under the most favourable circumstances, the compressed and saturated air would then be delivered to the cooling pipes at a temperature of at least 95 deg. Fahr., the pressure being say 65 lbs. per square inch absolute. Without taking into account any water mechanically suspended in the air, the quantity of aqueous vapour contained in it under these conditions would be 0·008 lb. to the pound weight of pure air. Now, as there is precisely the same weight of dry cold air circulating outside the cooling tubes in a given time as there is of warm compressed air within, it follows that, by whatever amount the temperature of the internal air is reduced, by an equal amount must that of the external air be raised. But in addition the internal air has vapour mixed with it, which, as the temperature falls, gives off heat, measured not only by the reduction in its sensible temperature, but by the latent heat of vaporisation; and this heat also has to be taken up by the external air.   It is found by calculation that, assuming each lb. of internal air, with its proportion of vapour, to be reduced to 42 deg. Fahr., the lb. of external cold air, which has to take up all the heat due to this reduction, will be raised in temperature by 84 deg. Fahr.

Instead of using the spent air for cooling purposes, the cold air from the expansion cylinder may be applied direct to the cooling apparatus; but in this case difficulty would be experienced from the deposited moisture inside the tubes actually freezing from the intense cold of the external air, a difficulty which the author understands has often occurred with this apparatus.   This, apart from the mere obstruction of the pipes, would involve a further sacrifice of cold, owing to the liberation of the heat of liquefaction.

It should, however, be stated that these machines have been worked successfully in cases where a large amount of cooling water at a low temperature is available, as, for instance, on board an ordinary Atlantic steamer.   There is no doubt that moderately dry air would be obtained wherever a sufficient supply of water at 45 deg. Fahr. or 50 deg. Fahr. can be had.

Sturgeon's refrigerator is a horizontal machine with some novel arrangements as regards the construction of its air valves and pistons.   The compressed air is first cooled partially, by being passed through tubes surrounded by cooling water; and is then passed through charcoal or some other absorbent of moisture, before being admitted to the expansion cylinder.

Messrs. Hick Hargreaves & Co. of Bolton manufacture cold-air machines of horizontal form, in which the Corliss cut-off gear is applied to the admission valves of the expansion cylinder. The air is compressed in a double-acting cylinder, into which cooling water is injected at each stroke; it then passes through a series of receivers, in which the water mechanically carried over is deposited; and it is finally admitted to the expansion cylinder, and expanded down to atmospheric pressure. So far as the author knows, no attempt is made at drying the air, which passes to the expansion cylinder fully saturated for its temperature and pressure; but a large snow box, consisting of a series of baffles, abstracts the bulk of the snow from the cooled air, after expansion and before its introduction to the cold chamber. In a machine of this description, we understand the snow has to be cleared out from the exhaust valves every few hours.

Messrs. Hall's cold-air machine is of both horizontal and vertical type, the latter applying to the smaller sizes, Fig. 34 being an illustration of one of their horizontal machines having a capacity of 70,000 cubic feet per hour. In either case, when combined with a steam engine, it consists of three double-acting cylinders placed side by side, at the end of a frame or bedplate, the cylinders are furnished with the usual moving parts, and the connecting rods work on three crank pins on a common crank shaft. One of the cylinders is used with steam in the ordinary manner, for giving the requisite motive power. Of the two others, one is for compressing, and one for expanding the air. The coolers are of

Fig. 34.

the multitubular type for surface cooling, and are placed below the bedplate or frame. The valves for the compression and expansion cylinders are slides of somewhat peculiar design worked from a pair of weigh bars, one for the main and the other for the expansion slides, the air-expansion cylinder having independent valves for inlet and outlet. The valves are placed on the upper side of the cylinders, which renders them very convenient of access, the design being such as to render all parts of the machine accessible, a most important element. The compressor is water-jacketed, and special steps are taken to dry the compressed air by means of a patent centrifugal moisture separator.

The Haslam dry-air refrigerator is also made both horizontal and vertical, the horizontal type applying to large machines and the vertical to those of small size. The cylinders are double-acting, and their arrangement with regard to one another varies in different classes of machines. The compressor is water-jacketed, and discharges into surface coolers placed in the bed. The compressed air, after having been cooled in the ordinary way by water, is further reduced in temperature in an interchanger, by the action either of the spent cold air on its way from the chamber in which it has been utilised, or of the cold air as it leaves the expansion cylinder; and in this manner a further condensation and deposition of moisture are brought about. The expansion cylinder presents no peculiarity in design, with the exception of the exhaust valves, which are separate from those admitting the air, and are so arranged as to offer as little obstruction as possible to the passage of the air.

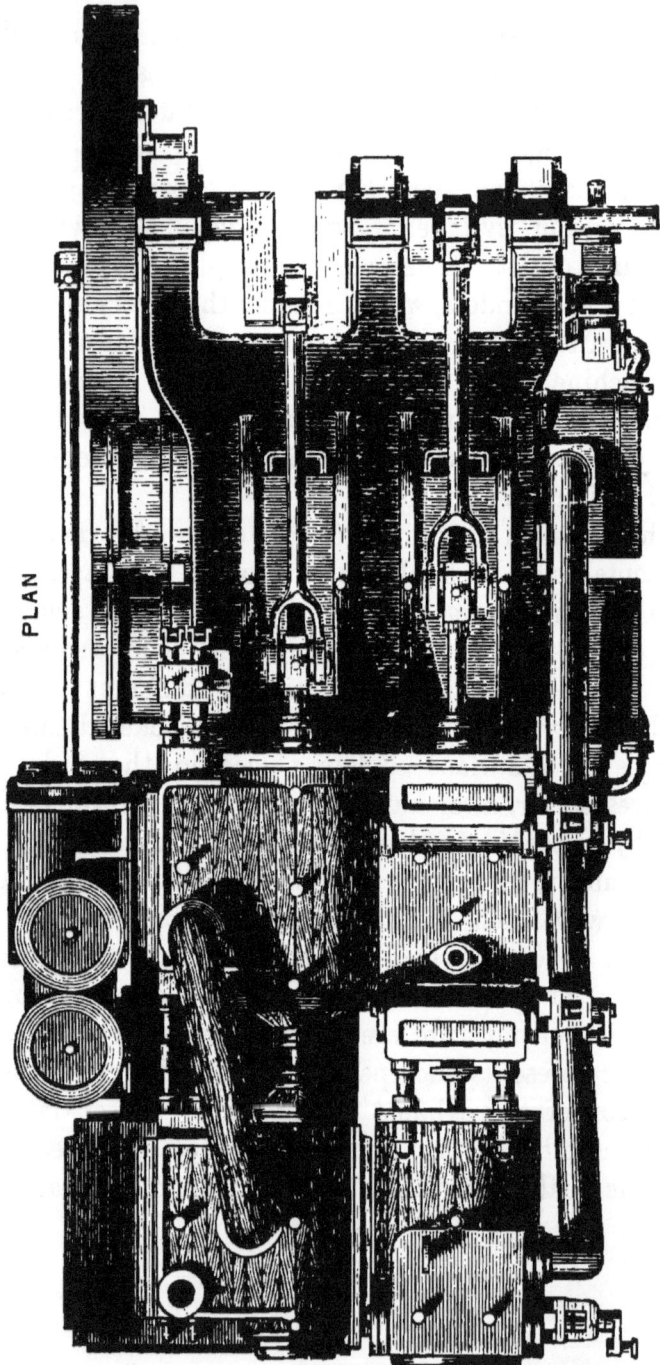

PLAN

Fig. 35.

In Figs. 35 and 36 is shown a horizontal dry-air refrigerator of Mr. Lightfoot's design, combined with a compound surface-condensing steam engine, of the type used for delivering from 60,000 to 120,000 cubic feet of cold air per hour. The compressor is double-acting, and expansion cylinder single-acting. They are placed close together, tandem fashion, with one rod common to both cylinders, the compression cylinder being nearest the connecting rod, leaving room for examination of the piston. In this way the coldest part of the expansion cylinder is removed from the hottest part of the compressor. The air valves are circular slides of phosphor bronze, actuated by eccentrics in the usual way. This kind of valve enables the ports to be made very short and direct; and besides being noiseless in action, it allows of a high piston speed being attained. The air enters the compressor through the valves at the top, and, after being compressed, passes by the pipe to the coolers, which are placed in the bedplate, and consist of a couple of iron vessels containing clusters of solid-drawn Muntz - metal tubes $\frac{3}{4}$ inch external diameter. Water is circulated through the inside of the tubes by the pipe (shown in Fig. 36), the supply passing in by the pump (also shown) through the tubes, and away by another pipe to the compressor jacket, whence it escapes. The water condensed and deposited from the air in the coolers is blown off from time to time by means of drain cocks, or may be discharged automatically. The compressed air passes through one cooler and returns through the second, being cooled to within some 5 or 6 degrees of the initial temperature of the cooling water, which circulates in a direction opposed to that of the air. The quantity

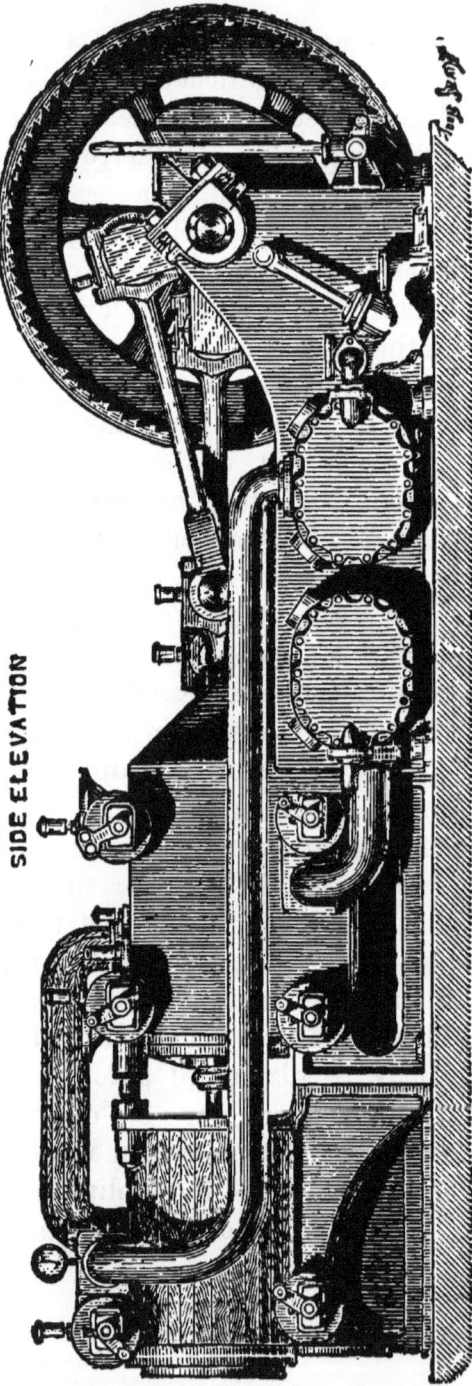

SIDE ELEVATION

Fig. 36

of water required is at the rate of from 30 to 40 gallons for every 1000 cubic feet of cold air discharged—that is, from three to four times the weight of the air; but the quantity varies in different machines, according to the efficiency of the apparatus. From the coolers the air passes by a pipe to the expansion cylinder; and after performing work upon the piston, and returning about 60 per cent. of the power expended in its compression, it is exhausted from the cylinder into the passage leading to the cold chamber, having been cooled down to from 70 to 90 degrees below zero Fahr. The steam cylinders are also placed tandem fashion when working on the compound principle. In this case the surface condenser is placed below the high-pressure cylinder, the air pumps being driven from a crank pin fixed in the flywheel. The same water serves both for cooling the air and for condensing the steam, passing first through the coolers and then through the surface condenser.

CHAPTER V

Comparison of the Various Systems—Carnot's Cycle—Maximum Efficiency of Machines—Cold-Air Machines—Ether Machines—Sulphur-Dioxide Machines—Ammonia - Compression Machines—Ammonia-Absorption Machines—Carbonic-Anhydride Machines.

THE various systems of refrigerating, together with the principal machines coming under each head, having now been fully described, it may be interesting to ascertain the comparative cost of producing ice by the various systems. In order to render this information as reliable and comprehensive as possible, the author considered it advisable to obtain the assistance of Mr. M. C. Bannister, C.E., and he takes this opportunity of expressing his indebtedness to him for his kindness and courtesy in furnishing the necessary calculations.

As we already know that refrigerating machines are governed by the laws of heat, and are, in fact, the reverse of an ordinary heat engine, it will be necessary to have a clear knowledge of an ideal heat engine and of Carnot's cycle as applied to it. According to Maxwell, a cycle may be described as follows:—A series of operations by which the substance is finally brought to the same state in all respects as at first is called a cycle of operations. For example, when the piston is rising the substance is giving out work; when it is sinking it is performing work on the substance which is to be reckoned negative. Hence, to find the work performed by the substance, we must subtract the negative work from the positive work. The remainder represents the useful work performed by the substance during the cycle of operations. If we have any difficulty in understanding how this amount of work can be obtained in a useful form during the working of the engine, we have only to suppose that the piston when it rises is employed in lifting weights, and that a portion of the weight lifted is employed to force the piston down again. As the pressure of the substance is less when the piston is sinking than when it is rising, it is plain that the engine can raise a greater weight than that which is required to complete the cycle of operations, so that on the whole there is a balance of useful work.

Let us now consider the relation between the heat supplied to an engine and the work done by it, as expressed in terms of the temperature.

If the temperature of the source of heat is T, and if H is the quantity of heat supplied to the engine at that temperature, then the work done by this heat depends entirely on the temperature of the refrigerator. Let T_2 be the temperature of the refrigerator, then the work done by H will be HC $(T - T_2)$. The quantity C depends only on the temperature T. It is called Carnot's Function of the temperature, which will be considered further on.

This, therefore, is a complete determination of the work done when the temperature of the source of heat is T. It depends only on Carnot's principle, and it is true, whether we admit the first law of thermo-dynamics or not.

10

If the temperature of the source is not T but T_1, we must consider what quantity of heat is represented. Calling this quantity of heat H_1, the work done by an engine working between the temperatures T_1 and T_2 is $W = HC (T_1 - T_2)$; and the heat being measured as mechanical work

$$H_1 = H - HC (T - T_1).$$

Now it is evident that if W equals the work done by the engine, and H_1 equals the quantity of heat supplied, then $\dfrac{W}{H}$ = the efficiency of the engine, but W was found to be equal to $HC (T_1 - T_2)$, and $H_1 = H - HC (T - T_1)$, therefore $\dfrac{W}{H_1} = \dfrac{HC (T_1 - T_2)}{H - HC (T - T_1)}$ dividing both sides by HC to eliminate these quantities the expression becomes $\dfrac{T_1 - T_2}{\dfrac{H - T + T_1}{HC}}$ or $\dfrac{T_1 - T_2}{\dfrac{1 + T_1 - T}{C}}$. On this theory, therefore, the efficiency of the engine working between T and T_2 is $\dfrac{W}{H_1} = \dfrac{T_1 - T_2}{\dfrac{1 + T_1 - T}{C}}$. It is plain that the work which a given quantity of heat H can perform in an engine can never be greater than the mechanical equivalent of that heat, though the colder the refrigerator the greater proportion of heat is converted into work. It is plain, therefore, that if we determine T_2 the temperature of the refrigerator, so as to make W the work mechanically equivalent to H_1, the heat received by the engine, we shall obtain an expression for a state of things in which the engine would convert the whole heat into work, and no body can possibly be at a lower temperature than the value thus assigned to T_2. Putting $W = H_1$ we find that $\dfrac{W}{H_1} = 1$ and therefore $\dfrac{T_1 - T_2}{1 + \dfrac{T_1 - T}{C}} = 1$ consequently

$$T_1 - T_2 = \frac{1}{C} + T_1 - T \text{ and } - T_2 = \frac{1}{C} + T_1 - T - T_1$$

$$\therefore \, - T_2 = \frac{1}{C} - T \text{ changing all the signs we have}$$

$$T_2 = T - \frac{1}{C} \text{ This is the lowest temperature any body}$$

can have. Calling this temperature zero, we find

$$T_2 = 0 \therefore T_2 = T - \frac{1}{C} \therefore \, 0 = T - \frac{1}{C} \text{ consequently } T = \frac{1}{C}$$

or the temperature reckoned from absolute zero is the reciprocal of Carnot's function C. Substituting T for $\dfrac{1}{C}$ in the above we have

$$\frac{W}{H_1} = \frac{T_1 - T_2}{1 + \dfrac{T_1 - T}{C}} = \frac{T_1 - T_2}{T + T_1 - T} = \frac{T_1 - T_2}{T_1} = \text{efficiency.}$$

We may now express the efficiency of a reversible heat engine in terms of the absolute temperature T_1 of the source of heat and the absolute temperature T_2 of the refrigerator. If H is the quantity of heat supplied to the engine, and W is the quantity of work performed, both estimated in dynamical measure, we have the following relation :—

$$\text{efficiency} = \frac{W}{H} = \frac{T_1 - T_2}{T_1}$$

In reviewing the various processes, it may be well to consider the precise nature of the work performed in any refrigerating machine.

To do this, there is no better way of dealing with the matter than by considering Carnot's formula when it is reversed; which will easily prove that if such an engine works between the lower absolute temperature and the higher absolute temperature, and it receives at the lower temperature a quantity of heat H which it gives away at the higher temperature, a certain amount of mechanical work W is absorbed.

And, further, if the compression and expansion of the refrigerating medium be carried out adiabetically, while in the reception of heat, H_1 at the lower temperature T_2, and its delivery at the higher temperature T_1 takes place isothermally, no other engine working between the same temperatures T_2 and T_1, and carrying the heat H from the former to the latter level of temperature, is able to perform this process with a smaller expenditure of mechanical work W.

That is to say, in a theoretically perfect refrigerating machine, the working mediums must carry out exactly the reverse process of Carnot's heat engine; provided the extraction of heat takes place at the lower constant temperature T_2; and, according to the laws of conservation of energy, the relation exists:—

$$C = R + JW,$$

J being Joules mechanical equivalent of heat;

W the mechanical work expended;

R the number of heat units extracted from the refrigerator;

C the number of heat units put into the condenser.

The working medium will be in exactly the same state after having completed the four operations (adiabatic compression, isothermal compression, adiabatic expansion, and isothermal expansion) as it was at the commencement; thus the second law of thermo-dynamics must be fulfilled, and consequently the relations hold good.

$\dfrac{R}{C} = \dfrac{T_2}{T_1}$ or $C = R\,\dfrac{T_1}{T_2}$ Setting the two preceding values C

to one another, it follows $R + JW = \dfrac{T_1}{T_2}$ or $JW = \dfrac{T_1 - T_2}{T_2}\,R$

i.e. the efficiency of a perfect refrigerating machine, working between the absolute temperatures T_2 and T_1, may be

expressed by $\dfrac{R}{JW} = \dfrac{T_2}{T_1 - T_2}$ As the only condition in the

establishment of this efficiency consists in the operations taking place according to Carnot's reversed cycle, it

follows that the above rule for $\dfrac{R}{JW}$ applies to all machines

so working, irrespective of the nature of the medium employed, whether that medium be air, ether, sulphurous acid, carbonic acid, or ammonia, or any mixture of these.

The actual efficiency $\dfrac{R}{JW}$ of a refrigerating machine,

which is always less than the theoretical efficiency $\dfrac{T_2}{T_1 - T_2}$

will approach the maximum value $\dfrac{R}{JW}$ the more nearly

the higher and lower temperatures approach each other. That is to say, the smaller the difference between the condenser and the refrigerator the greater the efficiency of the machine. It is important that this rule should be carefully observed in designing such machines. It also points to extension of surfaces both in refrigerators and condensers, and a rapid circulation of the liquids and saturated vapours, to promote the transmission of heat through the metal containing the mediums. It also indicates that the heat of compression should not be raised higher than the temperature of the condensing water during compression, and that the pumps should be as small as possible, the clearances as small as practicable, with an absence of all valve pockets and chambers. The inlet orifices to the pumps should be of ample capacity; the valves should be arranged to open simultaneously with the movement of the piston on its return stroke, and should close immediately with the cessation of the piston. The discharge valves should be as light as it is possible to make them, and should be set to give a free delivery the moment the pressure in the cylinder attains the pressure due to the condensing water.

The most important features to be considered, either in designing or selecting a refrigerating system are as follows:—(1) The efficiency of the steam generator boiler; (2) the design, construction, and efficiency of the motor; (3) the design, construction, and efficiency of the vapour compressor; (4) the designs of the refrigerator and condenser, relative to the temperature of the condensing water, and the temperature at which refrigeration is desired.

Coming now to the comparative cost of producing ice by the various systems, cold-air machinery will be first considered. This class of machine has been extensively manufactured in the past; but on account of the consumption of coal, and constant trouble and annoyance caused by the accumulation of snow in the air tanks and valves (after long runs) and other circumstances, this class of machinery is being gradually superseded by more simple and inexpensive systems of chemical refrigeration.

Cold-air machines have been hitherto chiefly used in the preservation of perishable articles of food. Brewers and other manufacturers requiring cooling machinery have not taken kindly to them, chiefly for reasons of economy, the considerable space they occupy, together with the large boiler capacity required; further, the air being delivered in a vibrating condition, it is not suitable for controlling fermentation in brewers' tuns. Cold air, however, may be used, where fuel is not a consideration, for rapidly chilling hot meat; also for sinking wells, boring tunnels, etc., when water in contact with loose sand or gravel has to be dealt with.

The relative value of freezing by compressed air may be considered as follows :—

	Units.
One pound of water at 60° to water at 32° =	28·0 Fahr.
One pound of water at 32° to ice at 23°=9° × ·5 specific heat =	4·5
Latent heat of ice	142·0
	174·5
Losses in transfer, radiation, conduction, and absorption generally 37·5% =	65·2
Heat units per pound of ice	240·0°

Assuming the pressure of the compressed air at two atmospheres total, we can calculate the quantity of air necessary to freeze a pound of water, and the mechanical power required to do the work. One pound of air at one atmosphere and at 60 deg. compressed to two atmospheres, is heated 116 deg., and the specific heat of air where expansion is permitted being ·238, we have ·238 × 116 = 27·6 units of heat per pound of air; and to freeze a pound of water from 60 deg. requires therefore 170 ÷ 27·6 = 6·16 pounds of air, which, divided by ·0761, its specific volume, gives a total of 81 cubic feet of air at one atmosphere and at 60 deg.

To find the power required to compress this air, imagine an air pump 1 foot square and 1 foot stroke, thus holding 1 cubic foot, and let it discharge air compressed to two atmospheres, and heated 116 deg. by the compression, into a reservoir where it is cooled down by cold water to 60 deg. again. Let that compressed air be caused to pass through a refrigerator where it is allowed to return to its normal pressure, and in so doing to absorb from the water to be frozen the heat which it gave out when compressed, and let the pressure in the reservoir be maintained uniformly at two atmospheres.

The pressure on the piston of the air pump would be one atmosphere at the commencement of the stroke, rising to two atmospheres above a vacuum when the volume was reduced to ·6117 cubic foot, or when the piston had travelled 1 — ·6117 = ·3883 foot. Take the travel of the piston and the mean pressure the piston travels 1 — ·9346 = ·0654 foot with a mean pressure of (0 + 1·47) ÷ 2 = ·735 lb. per square inch requiring ·0654 × ·735 = ·048 foot-pound; the next travel is ·9346 — ·8536 = ·081 foot, against (1·47 + 3·67) ÷ 2 = 2·57 lbs. requiring ·081 × 2·57 = ·208 foot-pound, and so on for say five intervals of the stroke. We thus obtain the following numbers :—

Foot lbs.

$$1\cdot0000 - \cdot9346 \times 0\cdot00 + 1\cdot47 \div 2 = \cdot048$$
$$\cdot9346 - \cdot8536 \times 1\cdot47 + 3\cdot67 \div 2 = \cdot208$$
$$\cdot8536 - \cdot7501 \times 3\cdot67 + 7\cdot35 \div 2 = \cdot570$$
$$\cdot7501 - \cdot6724 \times 7\cdot35 + 11\cdot11 \div 2 = \cdot717$$
$$\cdot6724 - \cdot6117 \times 11\cdot11 + 14\cdot7 \div 2 = \cdot783$$

Total, $= 2\cdot326$

It will be noticed that these intervals are unequal, being calculated from a table of pressures and volumes to save time. They could of course be taken at equal distances of the stroke, the result in both cases being precisely the same.

The work required to compress 1 cubic foot of air from one atmosphere, and at 60 deg. Fahr. to two atmospheres is thus found to be 2·326 foot lbs. per square inch of piston, or 335 foot lbs. per square foot.

This now compressed and heated air has then to be delivered against the pressure of one atmosphere, or 14·7 lbs. on the square inch, into the cooler or condenser. This delivery of air requires a further expenditure of power equal to 14·7 × 144 = 2116·8 lbs. per square foot, and as a cubic foot of air after compression is only ·6117 of its original volume then 2116·8 × ·6117 = 1295 foot lbs., which added to 335 foot lbs., the work expended in compressing the air, equals 1295 + 335, making a total expenditure of 1630 foot lbs. But by expanding this compressed air behind a piston, thus making it do mechanical work, there is a recovery of power

$= 2\cdot 326 \times 144 = 335$ foot lbs. -1295 foot lbs. $= 960$ foot lbs. during expansion, leaving a net expenditure of $1630 - 960 = 670$ foot lbs., which represents the work done in compressing and delivering 1 cubic foot of air taken in at atmospheric pressure at 60 deg., and, as we have already shown, 81 cubic feet of air are necessary to freeze a pound of water, then 81×670, and adding 335 for 50 per cent. friction in the pumps, thus $81 \times (670 + 335) = 81405$ foot lbs. per lb. of ice; further, adding losses due to radiation, friction, and conduction, equivalent to 17 per cent., makes a total of 95243 foot lbs.; and as 423 British thermal units is the net equivalent mechanical power exerted by a steam engine for the combustion of 1 lb. of ordinary coal evaporating 8 lbs. of water to steam of 100 lbs. pressure and expanded on the engine $2\frac{1}{2}$ times, there are only 423 thermal units of energy available in the pumps, this is equivalent to 327250 foot lbs., which being divided by $95243 = 3\cdot 43$ lbs. of ice per lb. of coal.

The amount of condensing water required to carry off the heat of compression would be $6\cdot 16$ lbs. of air $\times 116$ deg. $\times \cdot 238$ (specific heat) $\times 3\cdot 43$ lbs. of ice made $= 583$ lbs. of water raised 1 deg.; and as this circulating water should not be heated above 5 deg. $\therefore \dfrac{583}{5}$ $= 116$ lbs. of water would be required, which, if divided by $3\cdot 43$ lbs. of ice, equals 31 lbs. of water per lb. of ice produced.

The general objections to this method of cooling are :—

1. It is expensive in first cost.

2. A lubricant being necessary in the compression

cylinder, the action of the heat on the lubricant imparts to the air the disagreeable odour of burnt oils.

3. It is expensive to maintain, and involves excessive friction and constant stoppage to remove snow and accumulation in the trunks and valves, etc.

4. The large quantity of condensing and circulating water required.

5. The excessive boiler power, and consequent high wear and tear, and establishment charges.

6. Cold air being generally used direct, there is no reserve of cold that can be utilised during stoppage for · repair, should an accident occur to the machine.

7. The considerable amount of room occupied by the machine compared with the duty obtained.

On the other hand, the air process for artificial cooling has some advantages, as follows :—

1. It involves no serious difficulty in construction.

2. It is worked without chemicals, and where coal or power is of little consequence, it may be desirable to employ it.

3. For rapidly producing very low temperatures it is, though an expensive, still a most efficient process.

The second system to be considered is that of machinery for the abstraction of heat by the evaporation in vacuo of a more or less volatile liquid.

Sulphuric and methylic ether have been the liquids most commonly used for this class of machinery. James Harrison of Geelong, in the colony of Victoria, being one of the first to apply the system practically, and it is of considerable credit to him that the machines now in use are but slightly different from his original design, except in details of construction, etc.

From 1870 to 1880, ether machines had a large share of patronage, the principal firms supplying them being Messrs. Siddeley & Co., Messrs. H. J. West, and Messrs. Siebe, Gorman, & Co.   Messrs. Siddeley & Co. alone have made upwards of 500 or 600 of these machines; these latter were the first to be introduced into India on any large scale, and they are still holding their own and performing good service.

Ether may be described as a light volatile fluid, made from the distillation of alcohol with, generally, sulphuric acid.   It is highly inflammable, and only explosive when confined.   At atmospheric pressure its boiling point is 96 deg. Fahr.; the specific gravity of the liquid is ·72 at 60 deg. Fahr.; specific gravity of vapour, 2·58 (air being 1); latent heat of vapour, by weight 165 units, and by volume 324.   The latent heat at, say, ice-making temperature, or 20 deg. Fahr., and at $2\frac{1}{2}$ lbs. of pressure $= 197\cdot2$.

A complete ether machine, as already mentioned, comprises a vapour pump, and engine to drive it; the compression pump is generally coupled to the end of the piston rod of the engine, tandem fashion; a refrigerator which contains the liquid to be evaporated, and in which cold is produced; together with an ether cooler (which is supplied by the best makers), through which the vapour passes to the pump and cools the liquor as it returns to the refrigerator; a surface condenser to carry off the heat of vaporisation, and to liquify the vapour; a brine pump and water pump; and, if for ice-making, tanks, moulds, agitating gear, traveller, and hoist.

In describing the refrigerating machine, as we have just seen, there are 423 thermal units available in power for actuating the compressor.

This is the thermal dynamic equivalent for the compression of 11·71 lbs. of ether at a pressure due to 20 deg. Fahr. As before stated, the latent heat of ether is 165 thermal units at atmospheric pressure; but at 2½ lbs. absolute pressure the boiling point is lowered to 20 deg. Fahr., and the latent heat increases to 197·2 thermal units, consequently the vaporisation of 11·71 lbs. of ether would require 11·71 × 197·2 = 2309·21 units of heat. This heat is taken partly from the ether, which had an original temperature of 60 deg. Fahr., and partly from the brine or other substance in contact with the refrigerator. The first vaporisation would cool the ether from 65 deg. Fahr., or 5 deg. above the condensing water temperature, to 20 deg., the boiling point, at working pressure, or 45 deg. in all; taking the specific heat of ether as ·47, this requires 45 deg. × 11·71 × ·47 = 247·66 thermal units—there is also the loss due to the friction of the piston and the rod of the compressor, which in a well-designed pump absorbs about 10 per cent., also a loss due to absorption, radiation, and distribution, and the usual losses from mechanical construction, such as clearances, valve pockets, heat of pump, differences of temperature of compressor and refrigerator, which are equal to a further 7 per cent., in all 17 per cent., or $\dfrac{2309·21 \times 17}{100} \% = 392·56$ units, to which has to be added the difference due to the temperature of condensing water 247·66 units, making in all 247·66 + 392·56 = 640·22 units to be deducted; this leaves a balance of 2309·21 − 640·22 = 1669 units, which is the full beneficial results obtained in the refrigerator from 11·71 lbs. of ether for the expenditure of 1 lb. of good steam coal.

Then $1669 + 392\cdot56 = 2061\cdot56$ thermal units to be extracted and carried off by the condensing water; this multiplied by the specific heat of the ether, $\cdot47$, and divided by 5 deg. will equal $\dfrac{2061\cdot56 \times \cdot47}{5} = 193\cdot76$ lbs. of water heated 5 deg. Fahr. for the production of $6\cdot7$ lbs. of ice made. The quantity of ice made may be found as follows: the thermal units carried off by the condensing water multiplied by the specific heat of ether and divided by the latent heat of ice 144 deg., thus $\dfrac{2061\cdot56 \times \cdot47}{144} = 6\cdot7$ lbs. as above.

The compression of the vapour taken into the pump at a tension of $2\cdot25$ lbs. per square inch, or at its equivalent temperature of 20 deg. Fahr., and discharged at a tension equivalent to 5 deg. Fahr. above that of the condensing water, say 65 deg. or $8\cdot2$ lbs. per square inch, and at a volume due to this pressure, will require in power an expenditure equivalent to $423\cdot9$ thermal units or $423\cdot9 \times 772 = 327,250$ foot lbs. or the net equivalent mechanical power exerted by the engine from the combustion of 1 lb. of ordinary coal, to 8 lbs. of steam at 100 lbs. absolute pressure, expanded $2\cdot5$ times.

The third system comes under the heading of machinery for the purpose of expanding and compressing condensable gases.

During 1880, Professor Raoul Pictet of Geneva took out letters patent for the application of sulphur dioxide as a refrigerating agent, and has since introduced, more or less successfully, a number of machines, most of which, however, have been erected on the Continent. Sulphur dioxide is a heavy but volatile liquid at 25 lbs. pressure on the square inch absolute; it is a compound gas, composed of 1 part of sulphur and 2 parts of oxygen. It is usually prepared by heating sulphuric acid with copper; it is very soluble in water. The acid has great power of bleaching vegetable substances, and is not a supporter of combustion, nor is it combustible; water will dissolve 50 times its own volume of sulphur dioxide. It has a latent heat of 182 units at atmospheric pressure, and a specific heat of ·34 at constant pressure, and ·279 at constant volume, air being 1 for equal volume, and relatively to water ·15 and ·12, a specific gravity of 2·24; the volume of 1 lb. weight at 32 deg. Fahr. and at atmospheric pressure equals 5·5 cubic feet. The liquid boils at 14 deg. Fahr. under atmospheric pressure, and at −105 deg. Fahr. can be solidified. The latent heat at ice-making temperature and at 20 deg. = 181 units; the volume = 9·36 cubic feet at the same temperature and pressure.

A complete sulphurous acid machine includes a compressor and an engine to drive it, usually direct, as in the ether machine, from the end of the piston rod; a refrigerator and a condenser containing the liquid to be expanded, and a liquid meter. The refrigerator of this machinery is usually placed in the brine tank, when ice-making is the object, and consists of a number of serpentine coils, connected at their ends to two large pipes, one at the top and one at the bottom, which are again united by similar vertical tubes. Professor Pictet allows the acid to flow into the refrigerator at one end— the refrigerator being surrounded by brine, the acid immediately boils and expands into vapour, which is then carried off by the compressor through suitable connecting pipes and compressed into the condenser at the pressure due to the temperature of the condensing water.

Professor Raoul Pictet allows his gas on expanding in the refrigerating tubes to find its own way through the serpentine coils, at the same time having a through connection with the pipe to the compressor; the consequence is that the vapour finds its way to the point of least resistance, and never, if it can help it, traverses the coils; thus producing waste of surface and loss of efficiency.

There is also the brine pump, a water pump, and, if for ice-making, similar gear and arrangements as in the ether process.

As stated, the latent heat of the liquid is 182 units at atmospheric pressure, but at 8·45 lbs. pressure absolute, the boiling point is raised to 20 deg. Fahr., and the latent heat decreased to 181 units.

The vaporisation of 13·4 lbs. of sulphurous acid requires $13·4 \times 182 = 2438·8$ units of heat; part of this heat is taken from the liquid which has a temperature of 65 deg. Fahr., and part from the brine in contact with the refrigerator; deducting the loss from the difference due to the temperature of the condensing water and the brine (which would equal 65 deg. $-$ 20 deg. $=$ 45 deg., and 45 deg. $\times$ 13·4 $\times$ ·1511 specific heat $=$ 91·12 units), together with 17 per cent. loss through radiation, friction, absorption, and heat generated by compression,

$$\text{thus :} \quad \frac{2443·5 \times 17}{100} = 415·39 + 91·12$$

$= 506·5$ units, there is a balance of $2443·5 - 506·5 =$ 1937 thermal units to be taken from the brine; which is the full beneficial result obtained from the expansion of 13·4 lbs. of acid expanded from 65 deg. Fahr., and equal to $\dfrac{1937}{240} = 8\cdot07$ lbs. of ice made.

The compression of the vapour taken in at the pump, at a pressure due to the temperature of the brine (20 deg. Fahr.), and discharged at the pressure due to the temperature of the condensing water (or 5 deg. Fahr. higher), will equal 60 lbs. to the square inch absolute, and at a volume due to the pressure of 8·45 lbs. to the square inch, will require in power 327,250 foot lbs., or 423·9 thermal units, which is the equivalent of work done by a high-class engine for the consumption of 1 lb. of coal by an evaporation of 8 lbs. of water per lb. of coal, to steam of 100 lbs. pressure, expanded 2·5 times, being an equivalent in production of 8·07 lbs. of ice per lb. of coal.

In 1870, Professor Linde, of Berlin, patented certain arrangements in which pure anhydrous ammonia could be utilised as a refrigerating agent, by means of expansion and mechanical compression. Since that date, Professor Linde has persevered in perfecting his patents and improving his system, and has now placed on the market a machine which, in point of high-class construction and efficiency in duty and arrangement, both for ice-making and cooling purposes, occupies a prominent position. Since Professor Linde first brought out his ammonia machine several other firms have adopted the same principle, and in the United States the manufacture of compression ammonia machines has become a very considerable industry.

There are several arrangements for constructing these

11

machines, each maker having his own idea as to the most efficient principle.

The principal feature of the Linde compressor is the temperature of the pump during compression, and this is a most important point. By means of a suitable arrangement, a small portion of liquid ammonia is carried into the pump with the vapour, at the commencement of each stroke, sufficient to cool it down to a refrigerating temperature, and so prevent excessive heating of the pump and undue friction, the gas being discharged from the pump at a little above the normal temperature of the condensing water, instead of a temperature of compression, usually in other machines something over 180 deg. to 220 deg. Fahr., at 130 lbs. absolute pressure. This heat has, of course, to be got rid of at the expense of a portion of the work of the machine—in other words, the cylinder has to be proportionally slightly larger; but against this loss has to be considered the fact that there is no circulating water required, no water jacket; the friction of piston and rod is comparatively nil, and the wear and tear is considerably reduced; the gas and oil is never heated beyond the normal temperature in the pump, and consequently separates much more easily, and remains in good condition longer; and therefore this machine requires less attention, and will cost less for repairs under continuous work than any other make.

If Carnot's reversed cycle be considered for a moment, it is obvious that this machine is the only one carrying out the cycle of compression in a nearly perfect manner, as sensible heat of compression is strictly avoided during the period of compression.

In other ammonia compressors a great quantity of heat

is generated, and passes away through the delivery pipe of the machine into the water jacket of the compressor.

It may be pointed out here that no doubt indicator diagrams obtained from most other machines may appear to indicate more perfect results than those of this machine; but the fact must be considered that in such diagrams, on account of the increment of temperature, the compression curve rises a great deal faster than is necessary, and consequently the expenditure of work per stroke of compression is considerably greater than is requisite; the rapid rise of the curve indicates so much useless work converted into heat.

Ammonia is a compound, highly-volatile, condensable gas, having a boiling point of -37 deg. Fahr. at atmospheric pressure; a vapour pressure of 120 lbs. per square inch, at a temperature of 60 deg. Fahr.; a specific gravity of ·59 (air being 1); a specific gravity of liquid of ·76 at a temperature of 40 deg. Fahr.;—the latent heat at atmospheric pressure being 601 units; specific heat at equal weight, 0·5 ; at equal volume, 0·29 (water being 1).

The volume of 1 lb. weight at 30 deg. Fahr. and at atmospheric pressure is 21·9 cubic feet, and at a pressure due to the temperature of 20 deg. Fahr. will equal 41 lbs. per square inch, and will have a latent heat of evaporation of 582 units.

The vaporisation of 6·44 lbs. of anhydrous ammonia requires $582 \times 6·44 = 3748$ units of heat. Part of this heat is taken from the liquid, which has a temperature of 65 deg. Fahr., and part from the brine in contact with the refrigerator. Deducting this loss, which is equivalent to 129 units, and deducting 17 per cent. loss through radiation, friction, absorption, and heat generated by compression, there is a balance of $3746 - 129 = 3617 - \dfrac{3617 \times 17}{100} = 3617 - 614{\cdot}89 = 3002{\cdot}11$ thermal units to be taken from the brine, which is the full beneficial result from the expansion of 6·44 lbs. of anhydrous ammonia. As 240 heat units are required per lb. of ice then $\dfrac{3002{\cdot}11}{240} = 12{\cdot}5$ lbs. of ice made.

The fourth system is that of machinery for the physical evaporation, compression, and expansion of gases, and their absorption by aqueous solution.

Ammonia, from its high solubility in water, is the only element that has been used under this system. The ammonia absorption machine is, to all intents and purposes, a compression machine; but, instead of having a mechanical compressor and steam engine for working it, the compression is carried out physically instead of mechanically.

This physical compression is a very fascinating study, and has occupied the minds of many eminent men for a long time in the endeavour to obtain advantages over mechanical compression without the complicated and wasteful method of obtaining power by means of a steam engine. At first sight, and without going into the matter thoroughly, it would appear that an absorption machine with physical compression would be considerably more economical than a mechanical compressor, as the latent heat of steam would be used in the one case and only the converted heat in the other; but unfortunately, so far, this prospect has remained unrealised.

Taking the work performed, and going through the operation of this system, it is found that 20 lbs. of liquid ammonia of commerce, of ·88 specific gravity, put into the generator, evaporated, condensed, and passed into the refrigerator ; then absorbed and returned to the generator again, requires heat or steam equivalent to 5591 units, and delivers into the refrigerator only 7·2 lbs. of concentrated ammonia. According to calculation, there is available from the steam, after deducting 35 per cent. for radiation, conduction, and distribution, only 5581 thermal units—which is the equivalent necessary for the evaporation of 20 lbs. of liquor ammonia, and, as only 4·4 lbs. of anhydrous ammonia are available for expansion in the refrigerator, equivalent to $582 \times 4·4 = 2560·8$ thermal units, and deducting the loss due to the temperature of the condensing water, and 17 per cent. for conduction, waste, etc., thus :—$45 \times ·5 \times 4·4 = 99$ and $\dfrac{2560·8 \times 17}{100} = 435·3$ and $435·3 + 99 = 534·3$ units lost; deducting this from 2560·8 leaves a balance of $2560·8 - 534·3 = 2026·5$ units, which divided by 240, thus $\dfrac{2026·5}{240}$, equal to 8·8 lbs. of ice made. This is the full beneficial result for the evaporation of 4·4 lbs. of anhydrous ammonia in the refrigerator for the combustion of 1 lb. of coal, having a boiler efficiency of 8 to 1. But this efficiency may be considerably increased by a more perfect process of separation, and Messrs. Siddeley & Company Limited state that they can now obtain absolutely pure anhydrous ammonia in their rectifier. If this be the case, a great advance has been made in the right direction; but in the opinion of those who have worked these machines, the separation in a perfect degree is almost an impossibility. Nor can it be expected, with constantly repeated vaporisation at high temperatures, that aqueous vapour will not pass over ; even with condensing water at 50 deg. Fahr. a certain amount of vapour must be taken over.

The carbonic-anhydride system has recently been placed in the market by Messrs. J. & E. Hall Limited, of London and Dartford, whose experience of some ten years' successful manufacture of the dry-air machine enabled them to bring to bear the information so gained in applying the chemical machine, with its very considerable economy, to the purposes which had been hitherto reserved specially for the dry-air machines.

The very remarkable record of this machine deserves some comment. Thus, in 1890, a pioneer steamer was fitted with machinery of this type for importing 1000 tons of frozen meat from the River Plate. The success attending this venture led to many imitators, and, at the present time, these machines in actual operation may be counted by hundreds, including more than fifty on board ship ; and the importation of meat by their means alone exceeds 50,000 tons per annum.

From the very limited data at hand for calculating the duty and efficiency of this agent, it is difficult to compile any exact tables ; but from the makers' statements it does not seem that any great advance, with regard to efficiency and economy, has been made by its introduction, but the advantages of this machine appear to consist more in the convenience of using a material which is at the same time a very powerful refrigerant, evaporating under atmospheric pressure at 125 deg. below zero Fahr., possessing no powerful fumes, and of so harmless a nature that, even were the whole contents of a machine to escape into the engine room, no harm would result to persons present ; for although carbonic-anhydride or carbonic-acid gas, if present in large quantities, will not support life, yet, the function of the human lungs being the conversion of oxygen into carbonic-acid gas, it is found that no inconvenience is felt when the air breathed contains the entire charge of a machine even of large capacity.

As carbonic acid does not attack any metals or materials, the parts as well as the joints in the machine can be made with those substances best adapted for the purpose.

To this should be added, that carbonic-acid gas is now obtainable nearly everywhere in the liquid form, being used for making aerated waters, raising bread, etc., and can be purchased for $2\frac{1}{2}$d. per pound.

The disadvantage of this refrigerating agent consists in the very heavy pressure required to liquefy it: thus, some of the parts of the machine are subject to 50 or 60 atmospheres, a pressure in common use for hydraulic-power appliances, such as lifts, cranes, etc. As, however, the machines consist almost entirely of tubes of very small diameter, all of which are tested to four times the pressure mentioned above, the machines cannot be considered less safe than an ordinary steam boiler, which will probably have been tested to twice the working pressure. As these machines require exceedingly small compressors, compared with machines using other chemicals, it follows that, in spite of the high pressure, the total load on the working parts is a very moderate one, so that the high pressure does not mean any excessive wear and tear.

Owing to the fact that an escape of gas from the

machine would not be followed by serious consequences, the makers fit each machine with a safety valve, consisting of a very thin sheet of copper, which possesses sufficient strength to withstand the normal pressure, but will give way should any neglect or ignorance of the attendant cause an undue pressure.   This peculiarity renders safe the small machines, which are generally in the hands of men who have no special knowledge of this or any other type of machinery.

The question of the value of carbonic acid as a refrigerating agent, when used in the tropics, where condensing water is necessarily at 85 or 90 deg. Fahr., is of great interest.   Notwithstanding the fact that at temperatures above 87 deg. the gas does not condense into a visible liquid, not only repeated experiment, but actual experience in numberless cases, has shown that there is no such falling off in efficiency as might be expected, seeing the absence of visible change of state between the gas before and after condensation.   Whatever, therefore, may be the physical condition after condensation of the material under tropical conditions, it is evident that it does not lose its latent heat of vaporisation, as it gives proof of approximately the same capacity for absorbing heat as it possessed at lower temperatures, when there is no doubt as to its change of state from the gaseous to the liquid on condensation.

Carbonic-acid machines have been actually at work for the last two years in India with cooling water between 85 and 90 deg. in several installations, and also in Burmah and Queensland under very similar conditions, and the reduced output of ice, due to the warm cooling water, is by no means excessive, not more, it is stated, than is found to take place with machines using other chemicals.

CHAPTER VI

Applications of the Various Systems—Extraction of Paraffin—Cooling and Ventilation—Cooling of Chocolate—Cooling Beer Wort—Fermenting Rooms—Beer Storage Rooms—Hog Cooling Rooms—Beef Chill Rooms—Insulation of Buildings—Comparative Value of Different Insulating Materials—Ice Manufacture—Expense of Manufacture—Ice Factory Buildings—Importation of Frozen Meat—Cold Chambers in Passenger Ships—Dairies and Dairy Produce—Breweries and Refrigeration.

IN considering the applications of the various systems, it is not intended to deal with the apparatus for abstracting heat by the rapid melting of a solid; inasmuch as, excepting for domestic purposes in localities where other ice is not available, its application is wholly special and very limited, being confined almost entirely to the laboratory. Nor, in regard to the machinery and apparatus for abstracting heat by the evaporation of a more or less volatile liquid, need much be said so far as ice-making and ordinary cooling are concerned. The various systems have already been explained in considerable detail, and sufficient information has been given upon which to base an estimate as to their economical application under any stated conditions. It is therefore chiefly with the machinery by which a gas is compressed, partially cooled while under compression, and further cooled by subsequent expansion in the performance of work, that the present considerations will deal.

Probably the earliest application of a refrigerating machine to manufacturing purposes was in 1861, when one of Harrison's ether machines was used by Mr. A. C. Kirk for the extraction of solid paraffin from shale oil. Since then, the manufacture of paraffin has been developed to a large extent, and at the present time there are very few works engaged in its production without a refrigerating machine of one kind or another. For the cooling of worts and of fermenting beer in breweries, refrigerating machines are largely employed. With English beer, which it is not necessary to cool below 50 deg. Fahr., the general practice is to reduce the temperature of the cooling liquor by passing it through the refrigerator of the machine, the cooled liquor being afterwards used in an ordinary brewer's refrigerator. For lager beer, however, which is fermented at about 40 deg. Fahr., the liquor is generally cooled by means of brine, and the temperature is brought down nearly to freezing point. The same machine is in this country frequently employed for circulating cooled brine through a series of pipes above the fermenting tuns, as well as for cooling the liquor; while, in lager-beer breweries, the whole of the fermenting rooms and stores are kept, the former at about 42 deg. Fahr., and the later at about 38 deg. Fahr., by means of cold brine circulating through pipes placed either on the ceiling or around the walls. For breweries, as well as for paraffin extraction, there can be no doubt that the most suitable machines to employ are those in which the cold is produced by the evaporation of a volatile liquid. Notwithstanding this, air refrigerating machines have been applied for both purposes in certain special cases, and have given good results, though at a larger expenditure of fuel. There are many instances, however, in which the extra cost of fuel may be more than counterbalanced by the advantages resulting from simplicity and compactness, and from the absence of all inflammable or corrosive chemicals. Besides this, the facility of application of cold - air machines is much beyond that of any other refrigerator. For these reasons they have been applied in dairies and in butterine works, in the latter case an additional advantage being gained from the rapidity with which the cooling can be accomplished, owing to the extremely low temperature at which the air is delivered from the machine.

The most extensive application of dry-air refrigerators, however, has been to the preservation of meat and other perishable goods. Although it had long been known that at low temperatures the decomposition of animal matter was arrested for an almost indefinite period, yet the practical realisation of preservation by cold was prevented from being carried out for want of a simple and efficient means of artificial refrigeration. The ·attempts that had been made to produce a refrigerated atmosphere by means of ice had not given satisfactory results, owing no doubt to the moist state of the air, which, cooled by contact with melting ice, was necessarily saturated, and brought about a musty taste and loss of flavour in the meat preserved in it. In 1878, however, upon the successful development of the cold-air machine, it became possible to produce a cold atmosphere, which, even at a temperature of from 35 deg. to 40 deg., never contained more than 50 to 60 per cent. of the moisture required to

saturate it.   Under this condition all danger from excess
of moisture as well as from excessive dryness was avoided ;
and the dry-air refrigerator was therefore speedily adopted
for preservative purposes.

Machines in which cold is produced by the evaporation
of a volatile liquid have also been applied for preserving
perishable foods.   This has been done, either by cooling
the rooms direct by means of overhead pipes through
which the cooled brine is circulated, or else by causing a
current of air from a fan to impinge against surfaces
cooled by an internal circulation of brine, and by then
passing the cooled air into the storage rooms.

As to whether the air machine or that employing a
volatile liquid is the best and most suitable, no general
rule can be laid down.   The simplicity, compactness, and
readiness of application of the former have secured it a
ready adoption in many cases where chemical machines
would have been wholly inadmissible ; but, on the other
hand, it is probable that air machines have frequently
been entirely misapplied.   For use on board ship there
can probably be no difference of opinion ; and nearly the
whole of the meat now imported into this country in a
cooled or frozen state is preserved by means of dry-air
refrigerators, while in only one or two cases is a portion
of it chilled and frozen on land by chemical machines.

The means adopted for the freezing and preservation
of meat are very simple.   They consist in lining the
room, or hold of the vessel, with material as impervious
to heat as practicable.   The construction of the lining is
altered in different cases, according to circumstances and
to fancy ; but it may be taken that an outer and an
inner layer of tongued and grooved boards, 1 inch or 1 ½

inches thick, with a 9-inch space between filled with charcoal, form a fairly good protection; while in some cases silicate cotton may be used with advantage instead of charcoal. A little extra care and expense bestowed on the insulation of a chamber are soon repaid; for, when the contents of the chamber are once reduced to the required temperature, the refrigerating machine has nothing further to do than to neutralise the heat passing through the walls: so that, the more perfect the insulation, the greater is the saving in fuel, in wear and tear of machinery, and in attendance. The cold air from the machine is usually admitted by ducts placed near the ceiling; and, after performing its cooling work, it is led back to the compressor, to be used over again, with the addition of a small amount of fresh air. In freezing a temperature of about 10 deg. Fahr., or even lower, should be maintained, and the carcasses should be hung so that the air can circulate freely around them. If, however, the meat has previously been frozen, as is generally the case with the cargoes brought from abroad, which are to a large extent frozen on shore, the carcasses are best packed as close together as possible, taking care to avoid injury through bruising, and to see that a free space is left for the cold air to circulate between the meat and the inner lining of the chamber. The temperature in this case need only be maintained low enough to leave a sufficient margin in case of the machinery having to be stopped for any slight adjustment or for oiling.

The capacity of a machine to be applied in any given case is determined by a consideration, firstly, of the cooling work to be performed on the material contained in the chamber; and, secondly, of the amount of heat that will pass into the chamber from without. With regard to the first, nothing need be said here. The second quantity depends upon the area of the walls, floor, and ceiling, their construction, and the difference between the minimum internal and the maximum external temperature. Experience has of course laid down certain general rules; but there are always special cases arising which require special treatment, and which can only be considered on the basis here set forth.

The trade in frozen meat has already necessitated the establishment of large stores, where the carcasses are received, and kept until they are required for consumption. A number of retail butchers also are now adopting cold stores of their own; and similar installations have been erected for poulterers, game dealers, and butter salesmen, but need not be further referred to.

In addition to the importation of dead meat, refrigerating machines have been applied for supplying fresh, cool air for the ventilation of ships' holds in which live cattle are carried. In this way a temperature of 100 deg. Fahr. has been reduced to 70 deg. in the height of summer, and the loss of cattle has been entirely prevented. No doubt the same system could be equally well applied for the cooling and ventilation of public buildings.

In 1882, dry-air refrigerators were first applied to the cooling of chocolate by Messrs. J. S. Fry & Sons, of Bristol, who adopted one of Mr. Lightfoot's horizontal machines with the double-expansion arrangement. Since then a number of similar machines have been applied for the same purpose in different parts of Europe and the United States, and works which had to be entirely

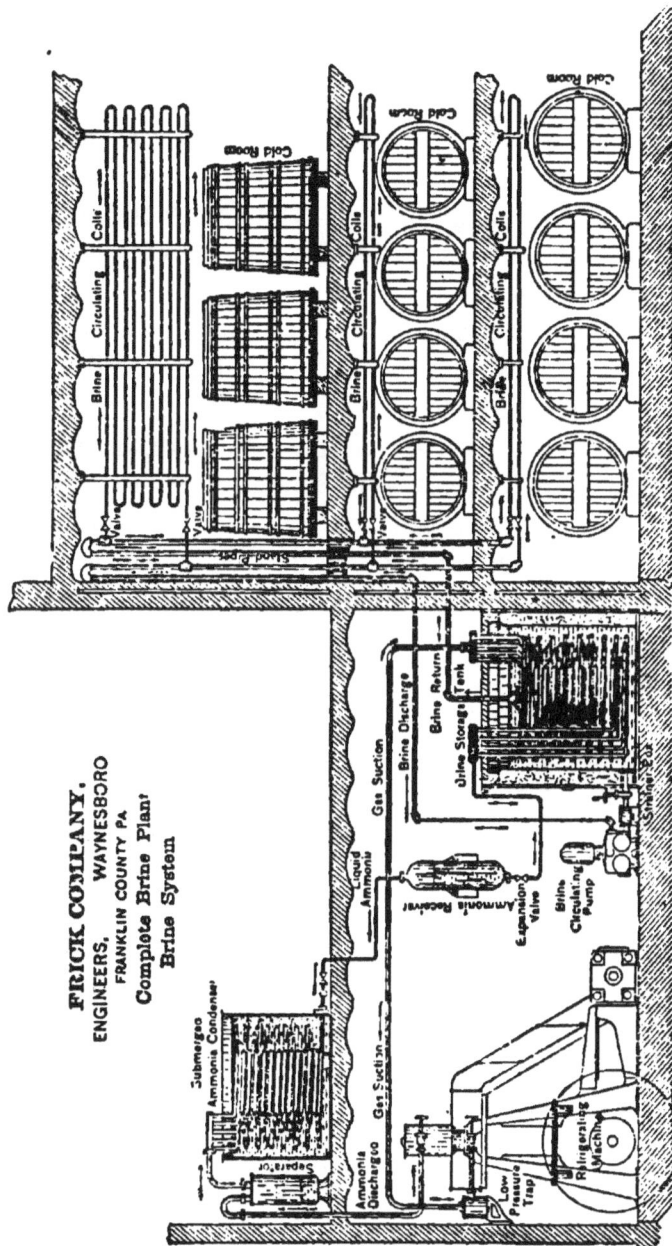

Fig. 37.

stopped in summer are now carried on during the whole year. The preservation of yeast, the cooling of gelatine dry plates and of fresh-killed meat in the slaughter-houses, and the freezing of tongues in South America for exportation, have all been satisfactorily accomplished by the dry-air machine.

The various applications of the modern method of cooling by use of mechanical refrigerating machinery of the compression type has in no instance of its widely developed fields of usefulness met with greater success than in breweries. It is a fact none the less interesting to know that the products of these breweries owe much of their fine quality, their growing popularity, and greatly increased demand to the introduction of mechanical refrigeration. This has also enabled the business to be carried on in the warmer latitudes where a very few years since it was considered impossible to operate a brewery, owing to the climatic conditions and great expense of producing natural ice.

One of the best known makers in the United States is the Frick Company of Waynesboro, Pennsylvania, who manufacture the "Eclipse" refrigerating machinery, to whom we are indebted for the following illustrations and for much of the information regarding refrigerating in American breweries and other establishments.

Fig. 37 is an illustration of their complete brine plant with pipes in rooms, and Fig. 38 their complete direct-expansion plant with pipe coils in position. As both these illustrations have the different parts carefully marked, and, in addition, arrows showing the direction of liquids along the various pipes, a further description of them will be unnecessary.

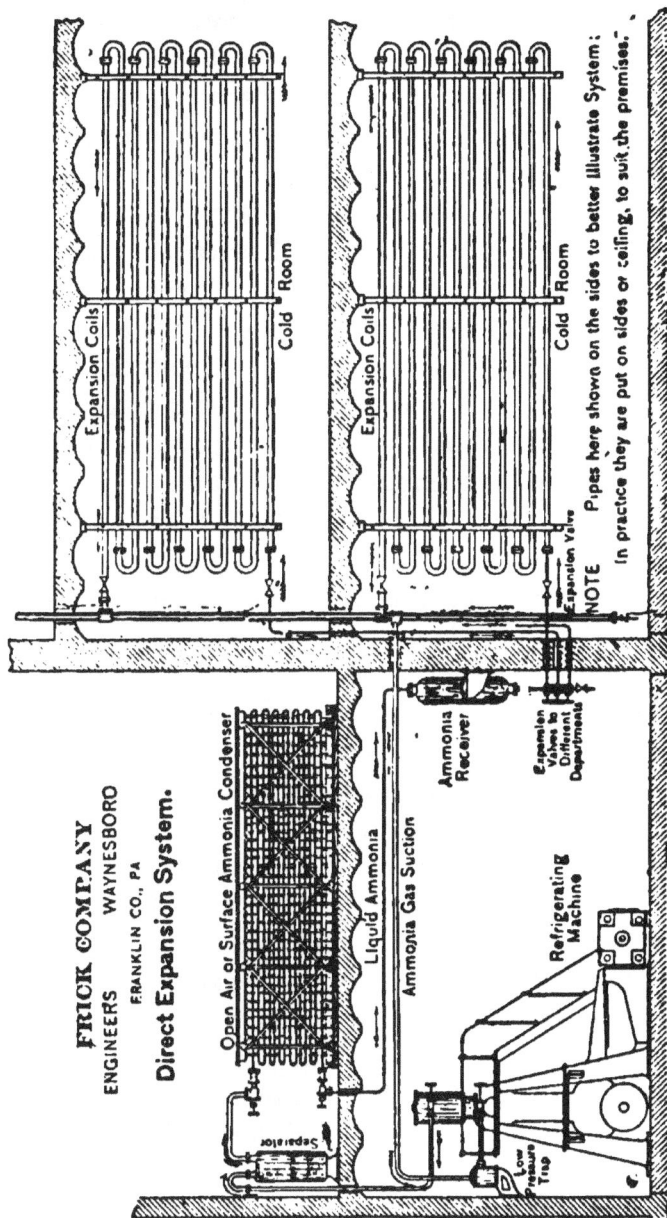

Fig. 38.

NOTE — Pipes here shown on the sides to better illustrate System; In practice they are put on sides or ceiling, to suit the premises.

FRICK COMPANY
ENGINEERS   WAYNESBORO
FRANKLIN CO., PA
Direct Expansion System.

One of the first operations in a brewery is the refrigerating of the hot beer wort which has previously been cooled somewhat by having been exposed in the "cool ship" (a large shallow tank generally placed upon the top of a building and housed in by open lattice-work, which allows the air to circulate and carry off considerable of the heat from the hot wort). Thence the wort flows down over what is called a "Baudelot" cooler, adapted for either brine or direct expansion, see Figs. 39 and 40, through the upper portion of which hydrant or well water is used, the lower portion of the cooler being mechanically refrigerated by direct expansion of ammonia, by brine circulation or circulating ice water.

The temperature of fermentation is regulated by attemporators through which ice-cold water or brine is pumped. In the brine system the ice-water is cooled in a cistern or suitable tank provided with either a direct ammonia expansion or brine coil, supplied by the refrigerating machine, the sweet or ice-water thus made being forced through the attemporator in the tubs, each or any number of which can be shut off or regulated at will, the pressure and amount of cooling water being under automatic control of the self-acting pump and regulator which supplies the attemporators and needs no attention, whether one tub or many be in use.

The introduction of mechanical refrigeration in packing houses and abattoirs marks a new era in this business, and has, as a matter of fact, brought about a revolution in the arrangement of buildings, and in the methods of caring for fresh killed hogs and beef and storage in packing houses.

As some of the older packing houses have not yet

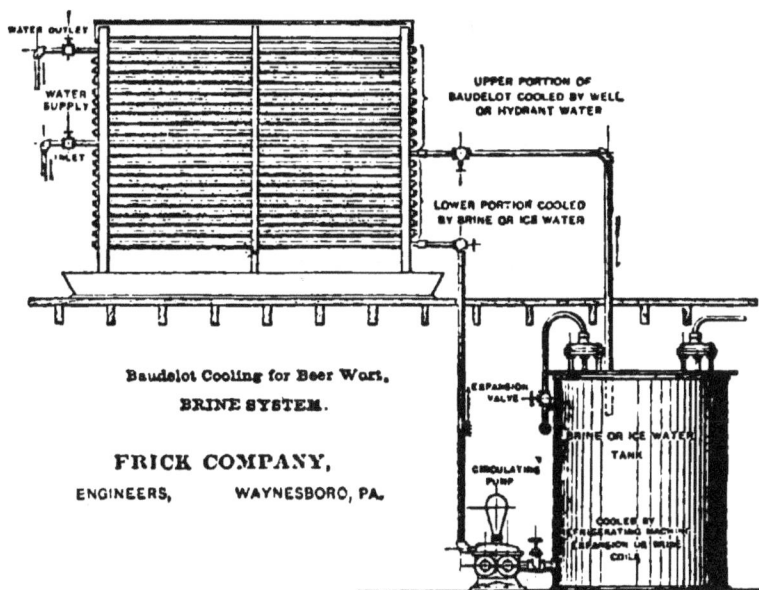

Baudelot Cooling for Beer Wort.

BRINE SYSTEM.

FRICK COMPANY,

ENGINEERS,   WAYNESBORO, PA.

Fig. 39.

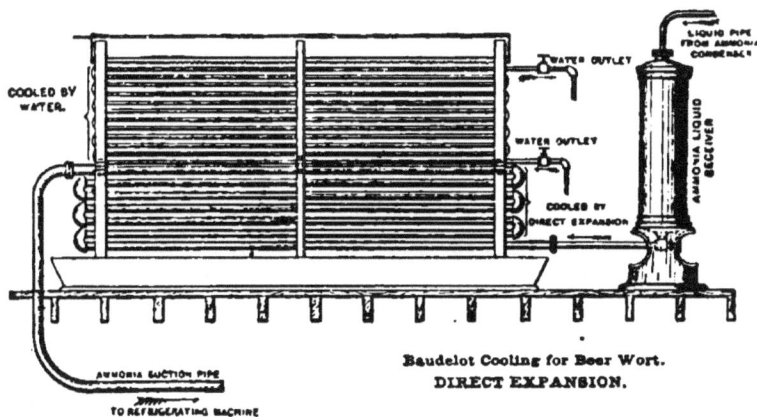

Baudelot Cooling for Beer Wort.

DIRECT EXPANSION.

Fig. 40.

adopted mechanical refrigeration, owing, perhaps, to difficulties in adapting a good system to their old buildings, such cases are met by a variety of methods, which can be introduced with little or no change in the premises.

By any of these plans of mechanical refrigeration, the temperature of a hanging or chill room can be regulated and maintained to exactly suit the requirements without regard to external conditions of the atmosphere and

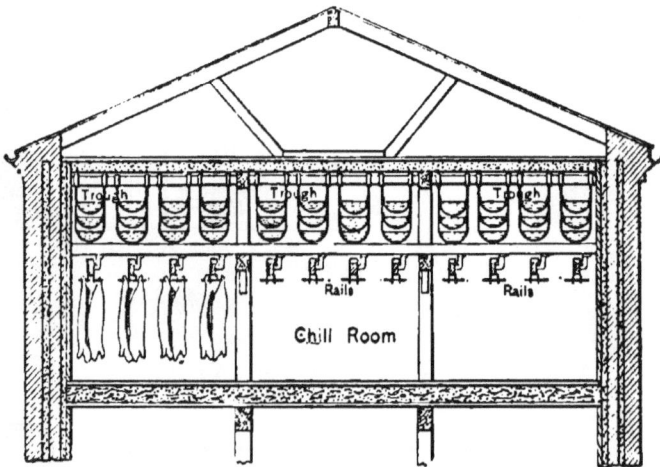

Fig. 41.

amount of internal cooling work to be done. There is no danger of such rapid freezing as to simply chill the outside surface of the carcasses and endanger the proper curing, because the animal heat can be taken up thoroughly by simply regulating the temperature, thus insuring the carcasses being uniformly chilled throughout.

The hog cooling room, Fig. 41, shows an adaptation of open cooling troughs over each hog rail, and is intended for brine circulation, the amount of brine flowing over the several surfaces being regulated by simple means. This plan presents advantages which will appear to those interested. Another plan for hog rooms, Fig. 42, is the use of coils in a cooling chamber overhead, with a proper construction and arrangement of air ducts introduced to insure ventilation and rapid circulation of the cold air. Still another plan for chill rooms, Fig. 43, is to place

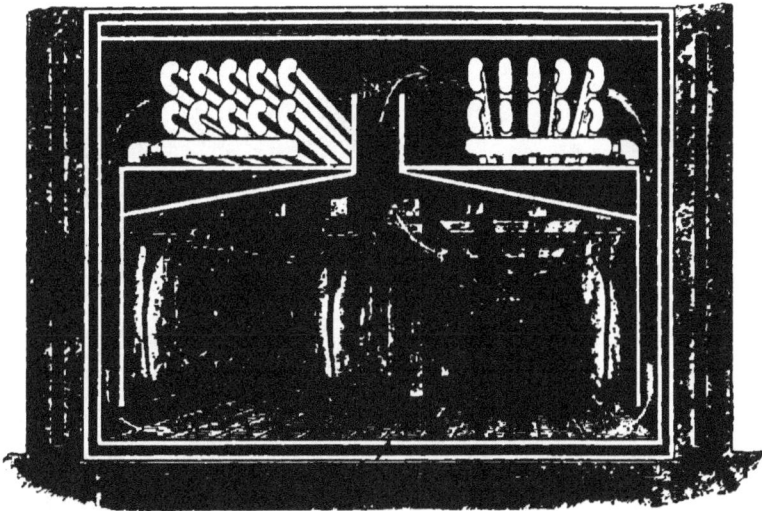

Fig. 42.

cooling pipes on the sides of the room, the hogs or beef being suspended between the coils.

For a beef room the plan shown in Fig. 44, with its system of large galvanised pipe and the simple arrangement of air ducts for cold-air circulation, is effective. As proved by extended use, it practically fills all the requirements, all drip and moisture being prevented. A beef room with box coils, Fig. 45, is another

favourite plan where room cannot be obtained for overhead cooling chamber.   All the above plans, except the
open trough system, can be used with either direct
expansion of ammonia or brine system.   –

The insulation of buildings used for the preservation
and storage of substances subjected to mechanical re-

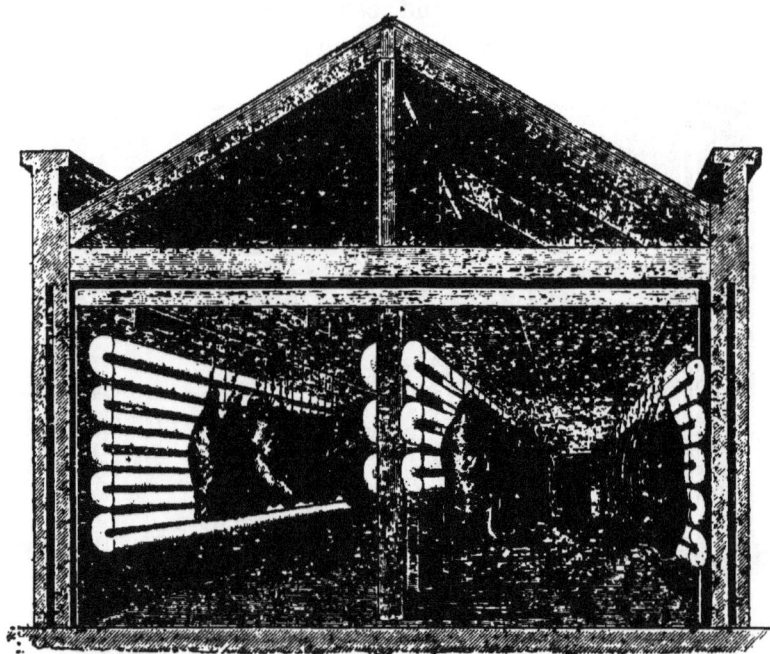

Fig. 43.

frigeration is a matter of vital importance when viewed
from an economic standpoint.   It is true that by
employment of a large surplus of refrigerating power,
errors of insulation, with its entailed great loss of
negative heat, is wastefully overcome and the desired
amount of cooling work accomplished; but this is a bad
way to reach a result, for the reason that it is a pre-

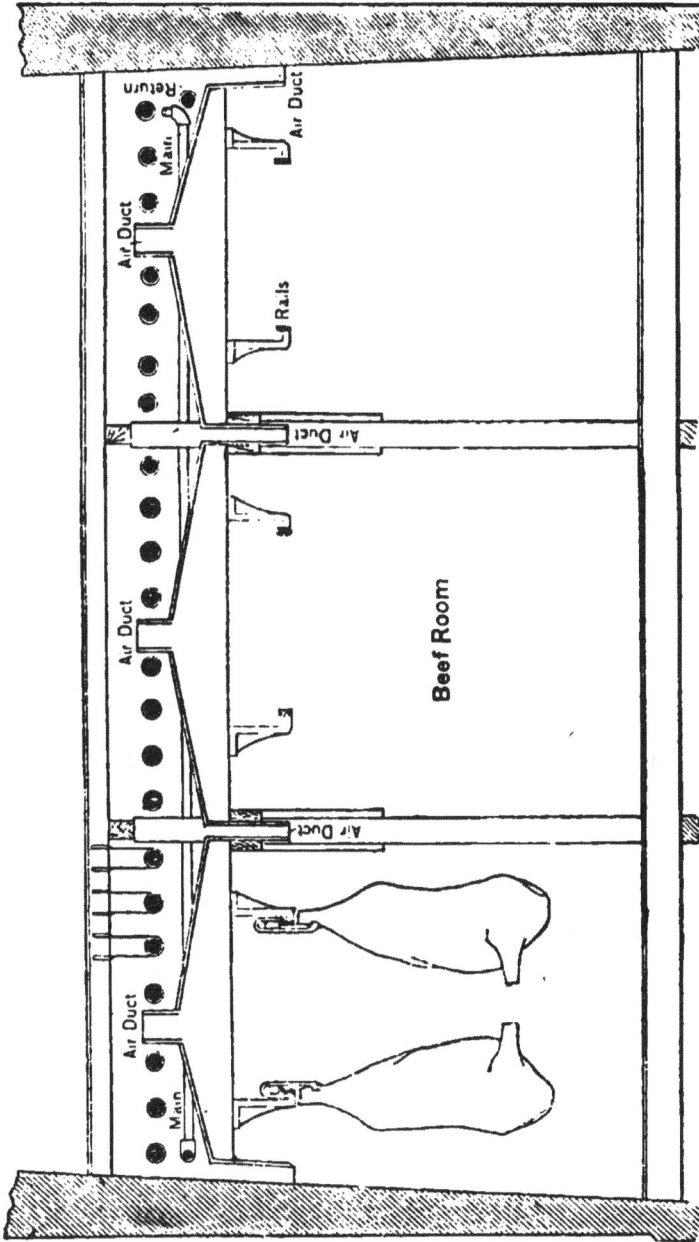

Fig. 44.

ventable loss.   Poor insulation is like paying interest on
borrowed capital, and a never-ceasing and useless drain
upon the machinery and pocket-book of the owner.

Perfect insulation absolutely prevents the transfer
of heat through the walls of a building; but this is
practically impossible.   If it were, one cooling of the
contents of the room would suffice; for there being no
loss, they would continue at the same temperature for an

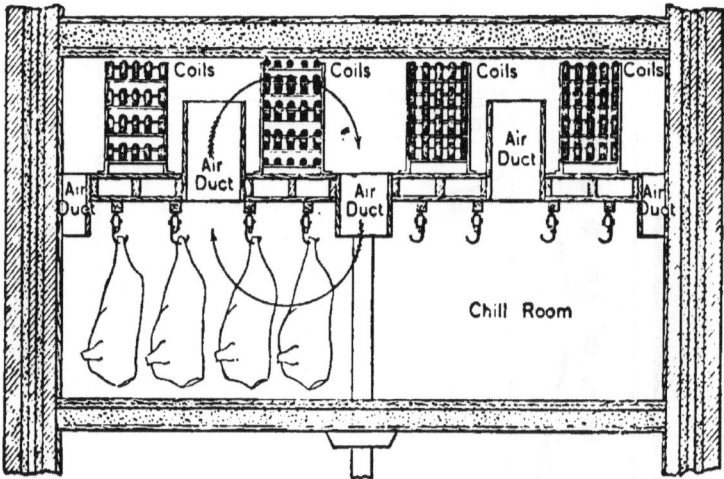

Fig. 45.

indefinite period.   If all articles placed in the room
thereafter were previously cooled to the temperature of
the room before placing therein, no work need be done
thereafter in the room itself.

A large percentage of the actual work of a refrigerating
machine is required to make up for transfer of heat
through the walls, floors, and ceilings occasioned by poor
insulation, and the amount may be experimentally determined by proper instruments.

Owing to difference in construction, exposure, and insulation of buildings, there is a great difference in economy of performance and work done by the same machine in use by different parties in the same line of business; and we here add, founded upon general experience and careful observation, that what a given machine and apparatus will do in one place is no certain guide for another place somewhat similar. The insulation, exposure, and method of handling the business is mainly responsible for the difference.

Several plans for insulation, all having been used with success, are shown in Figs. 46, 47, and 48.

The following is a table giving comparative value of different insulating materials:—

[NOTE.—This table gives the conducting power for each *square foot* of surface, and the comparative value is expressed in the number of *units of heat lost* by transmission through them *per hour*. They are arranged in the order of their merit.]

	Units Lost.		Units Lost.
Copper	515·	Brick dust	1·33
Iron	233·	Coke dust	1·29
Zinc	225·	Cork	1·15
Marble	28·	Chalk powder	0·87
Stone	17·	Charcoal powder	0·64
Glass	7·	Straw chopped	0·56
Brick work	5·	Coal dust	0·54
Plaster	4·	Hemp canvas	0·41
Double windows	3·6	Muslin	0·40
Oak wood	1·7	Writing paper	0·31
Walnut wood	0·8	Cotton	0·32
Pine wood	0·75	Air confined	0·3
Sawdust	0·55	Gray blotting paper	0·27
Indiarubber	1·37		

Tar paper, pitch, fine cinders, hair felt, etc., are used with success.

One of the conditions to be taken into account in dealing with low temperatures is the facility with which

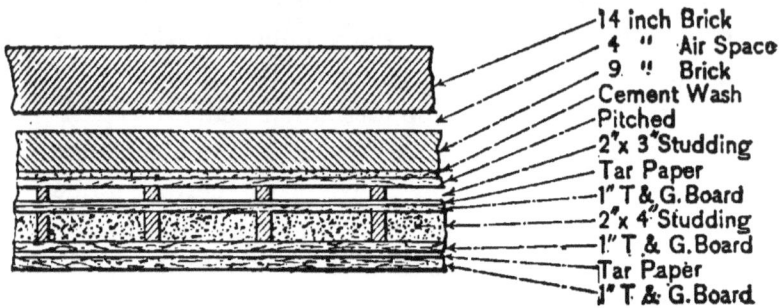

14 inch Brick
4 " Air Space
9 " Brick
Cement Wash
Pitched
2"x 3"Studding
Tar Paper
1" T & G. Board
2"x 4"Studding
1." T & G. Board
Tar Paper
1" T & G. Board

Fig. 46.

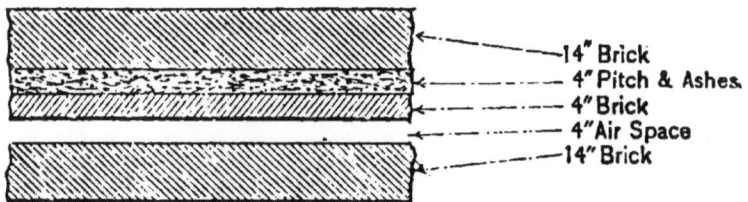

14" Brick
4" Pitch & Ashes
4" Brick
4" Air Space
14" Brick

Fig. 47.

36" Brick Wall
Pitch
1" Sheathing
4" Air Space
2"x 4"Studding
1" Sheathing
Mineral Wool
2"x 4"Studding
1" Sheathing

Fig. 48.

moisture from the air will be absorbed by some of the above substances, and lead to decay, softening, and acid fermentation. For this reason granulated cork, pitch, tar paper, wood well shellaced, or substances that will resist moisture, are the favourites. *Use only substances that have no strong odour or have same deodorised.*

The manufacture of ice upon a large scale is found to be profitable, and offers special inducements as a re- munerative investment for capital, for the reason that it can be produced so cheaply by the "Eclipse" and other first - class ice-making machinery that, except in the higher Northern latitudes, natural ice cannot compete with the machine product, either in cost or quality.

The water used for freezing in the ice moulds is chemically pure (distilled water), and of course free from all organic matters, disease germs, etc., which are seen in natural ice, and claimed by physicians to cause diphtheria, fevers, and kindred diseases. It is rare that natural ice can be found entirely free from impurities, as much of it must be gathered from filthy streams, stagnant ponds, canals, shallow and still waters, or basins into which sewers have emptied, or contaminated by receiving the surface washings of the soil, which is at all times laden with deleterious organic and decomposed animal matters.

The best natural ice is taken from the waters of the great Northern rivers and lakes, or large and deep bodies of water, which afford, by reason of the great severity of the winters, an immense harvest, and, under favourable conditions, clear crystal ice. Even this ice, under the microscope, is teeming with organic life, and is far from being as pure and wholesome as ice manufactured from distilled waters.

While Nature manufactures ice without cost to any- one, the cost of cutting, handling, transportation, loss by waste and meltage, and expense of storing when gathered from the ice-producing localities of the North, is so great that, even in cities as far north as New York, Philadelphia, and Chicago, manufactured ice can be produced and sold at a handsome profit, and in direct competition with the famous "Northern lake" and "Kennebec ice," as practically demonstrated by many ice dealers who now profitably operate ice factories.

An analysis of the reasons why ice can be manufactured and placed on the market cheaper than that gathered from natural sources will be interesting, aside from a consideration of the unquestioned superiority of the machine-made article.

Manufactured ice has one advantage among others, of being made in the very market in which it is sold and consumed, hence suffers no loss and occasions no extra expense for transportation and storage. The actual demand is supplied from day to day by making the ice as called for, drawn directly from the freezing moulds without waste, and supplied to the customers without intermediate storage and loss by meltage in transportation, which ranges from 20 per cent. upwards.

The operating expenses of an ice plant of a given capacity can be predetermined, as all expenses connected with the factory are fixed quantities for given rates of production. It is in every sense a routine business, with no greater contingencies and as susceptible of exact calculation as the business of pumping a given quantity of water.

The more extensive or the greater the capacity of the plant, the less the cost per ton of production, but even plants of a daily capacity of 5 tons per day prove remunerative investments.

Aside from the influence of the capacity of a plant, the cost of making ice also varies slightly with different localities, being affected by cost of fuel and labour, the difference being so little, however, that we find ice manufactured and sold in the South quite as cheaply as natural ice in the Northern cities. It is a matter of congratulation that the machine-made ice has supplanted natural ice wherever introduced.

In the application of a machine for making ice, it seems that this art may aptly be divided into two grand systems, the one using brine for the purpose of freezing water, the other effecting the freezing by direct expansion, the same as in refrigerating plants.

It is clear that it avoids a great deal of superfluous apparatus, of loss in efficiency and of untidiness, if the cooling or freezing is done directly without the interpolation of brine. Experimenters have for this reason tried to do away with brine in ice-making, as it has been abolished in the refrigeration of rooms. But here much greater obstacles have been met than in cooling plants. Given a good pipe system, and refrigeration by direct expansion has no difficulties whatsoever. Where, however, water has to be solidified, the first drawback met is the necessity of straight surfaces. A wrought-iron pipe of moderate diameter is the safest and cheapest means of confining ammoniacal gas, but if ice is formed around the pipes it becomes a matter of great wastefulness and trouble to loosen it again from the pipes. Straight surfaces are very difficult to construct and to keep tight, and all attempts to do this have been failures so far.

Another proposition to freeze water without the use of brine has been to imitate nature, viz.: to produce temperatures below the freezing point in well-insulated rooms. But here the low specific heat of air, and its low degree of conductibility, proved such a great obstacle that the cooling surfaces of the rooms had to be made excessively large, and still the result was extremely slow freezing. Still another form of ice machine is one which freezes the water *in vacuo*, without the use of either brine or any other agent than the water itself. If water is exposed to an almost absolute vacuum it turns rapidly into vapour, the transformation requiring so much heat, which is furnished by the water itself, that the balance of the water which is not vaporised freezes solid. The ice thus formed, however, is totally unfit for the market, being in the shape of granulated snow, full of air, brittle, and of no durability. This process was first proposed and introduced for the freezing of carafes by E. Carré (not the inventor of the absorption machine), and afterward carried out on a large scale by F. Windhausen in Germany in his vacuum ice machine. The system, however, was not a successful one, partly on account of the poor quality of the ice, partly because the sulphuric acid, which was used as an auxiliary to the air pump to carry away the aqueous vapour by absorption, caused great trouble in its process of reconcentration.

The small success which attended all the attempts at ice-making without brine has resulted in its being employed in all the different processes now in vogue.

In its application the system of making ice by the use of brine is quite varied, but, on the whole, three different modes have, up to the present, established themselves in the market :—

First, The system of removable cans.

Second, The plate system.

Third, The system of stationary cells.

The first is the one most in use the world over. In an iron or wooden tank, well insulated, a salt brine is kept at a temperature considerably below the freezing-point of water by evaporating coils, which are connected to the gas pump, if the machine is a compression machine, or to the absorber if the machine is an absorption machine. Fig. 49 shows a section of a can freezing tank, and discloses the interior arrangement thereof, showing the arrangement of ammonia evaporating pipes, ice moulds, and framework for holding the cans in position with wooden covers.

Parallel rows of ammonia pipes with space between each set to admit of a row of moulds being submerged in the cold brine with which the tank is filled, the water being frozen by the method described in chapter iii. The contents of the cans are frozen into a solid sparkling mass of ice in from 36 to 60 hours, depending upon thickness of mould and temperature of brine.

The ice moulds are made in various sizes, generally 100, 150, 200, and 300 lbs. The actual weight of cake is about 10 per cent. greater, however, to allow for wasteage. The demand of the locality influences the selection of the size of cake and determines the weight to be used.

A suitable hoisting arrangement, travelling upon iron rails, over each tank is used for lifting the moulds out of the tanks, and they are then carried by the hoist to a

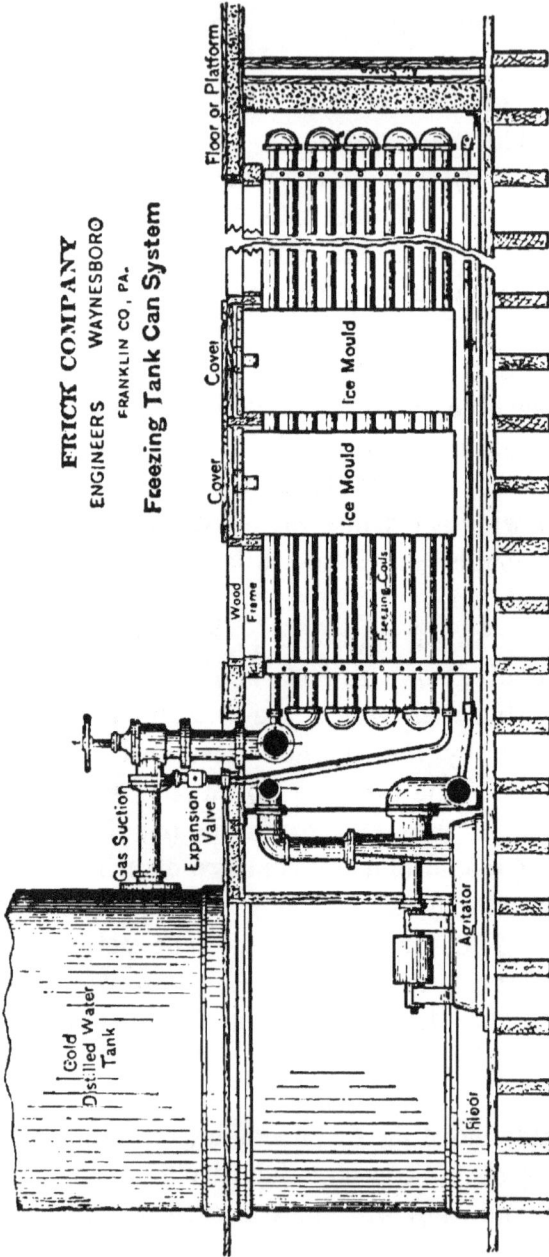

FRICK COMPANY
ENGINEERS   WAYNESBORO
FRANKLIN CO., PA.
Freezing Tank Can System

Floor or Platform

Cover

Cover

Ice Mould

Ice Mould

Wood Frame

Freezing Coils

Gas Suction

Expansion Valve

Cold Distilled Water Tank

Agitator

Floor

Sectional Side View

Fig. 49.

thawing device, Fig. 50, and immersed for a moment in water of say 50 to 70 deg. The mould is then tipped over an inclined runway, the cake of ice easily slipping out of the can and sliding down the runway through a trap into the ice house or anteroom, where it is temporarily stored, awaiting delivery to customers.

In the plate system, Fig. 51, which, as a rule, produces ice in pieces weighing one or more tons, a hollow plate of

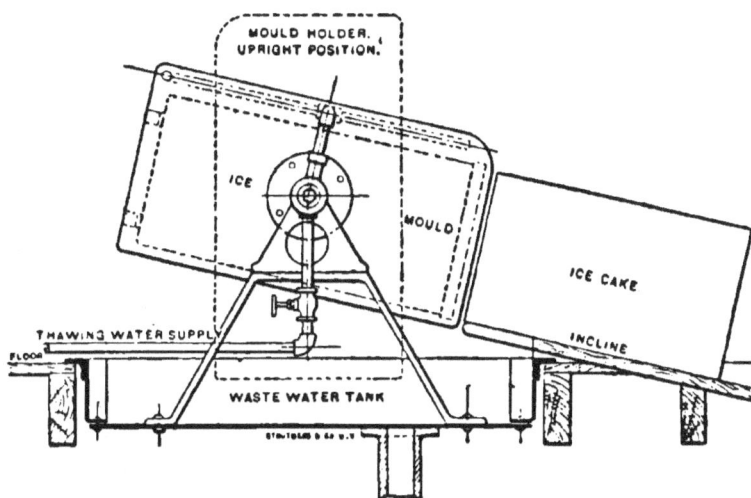

Fig. 50.

boiler iron is formed and immersed in a tank containing fresh water to be frozen. This plate is filled with brine, which is kept below the freezing point by evaporating coils in a manner similar to those of the can system. The coils may be either in the plates or outside in a separate brine tank, and the brine circulated through the plate. By thus keeping the plate at a sufficiently low temperature, ice will form on both sides of it, and by and by two layers of ice will be built up on the two sides of

13

the plate.   In order to remove this ice, the cold brine is drawn from the plates, and, in case the evaporating coils are inside of the plates, the circulation of ammonia in them is stopped.   Then tepid brine is supplied to the hollow plates, and after a short time the ice is loosened from them, and can be hoisted out of the tank by means of cranes, and cut up into blocks of any desired size.   A number of plates are as a rule immersed into each tank, and a whole tank emptied at one time.   In order to make the process continuous, more than one tank must be supplied, so that one at least is in continuous operation, while the other is being emptied and refilled and prepared again for work.   But on larger plants even more . than two tanks are necessary to permit of a daily drawing of ice.   The freezing process going on from one side only, *i.e.* a certain thickness of ice being formed by building up only on one side, the time of freezing is necessarily long.   In a can, ice is formed on two opposite sides, and the two surfaces growing together in the centre will ultimately make a solid block equal in thickness to the width of the can.   If ice of such thickness is made on a plate, frozen only from one side, it takes about four times as long.

In the system using stationary cells the cold brine is pumped through the hollow walls of the cells, the latter being open at the top, and filled nearly brimful with the fresh water to be frozen.   Ice will form in the cells the same as in the can system.   After the blocks are finished in the cells, tepid brine is pumped in place of the cold brine, and thereby the ice loosened from the cells, and its removal becomes a matter of little difficulty.   It is self-evident that in this system a whole tank has to be

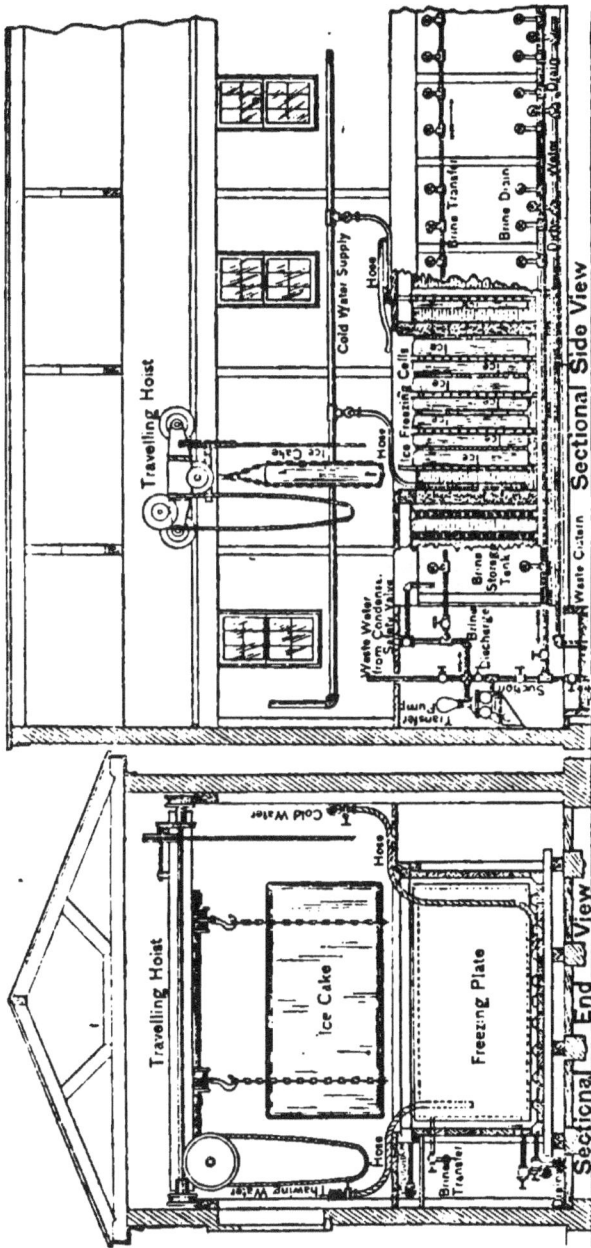

Fig. 51.

emptied at the same time as in the plate system, and, to make the plant continuous in its operation, more than one tank has to be employed. If the cells are made quite deep in proportion to their width, similar to the cans used in the can system, then of course the freezing-time is as fast as in the system first described. But if shallow cells, pan-shape, are used, the depth being small in proportion to length and width, then the freezing will practically be done mostly from the bottom, and for the same thickness of ice the time of freezing will be quadrupled as in the plate system.

In the beginning of the industry of ice-making, many manufacturers were satisfied with producing an article regardless of quality. Therefore no special pains were taken to make transparent ice, but by and by the demands for a better product were made. At first, freezing at comparatively high temperatures was resorted to, by which at least one part of the block became clear. But then, the time of freezing was so slow, and it took such a large number of cans and large tanks, and the first cost of the plant came to be so high, that means were tried to make the ice faster, freezing it at lower temperatures and still making it clear.

Quite a number of inventions were made to obtain this object, all of which were more or less successful. One thing was soon discovered: that clear ice could be produced by agitating the water during the process of freezing; the different propositions to accomplish this being quite numerous. A metal bar was let into the can and lifted up and down by a small revolving shaft and thumb, or a crank; or a wooden paddle was inserted into the can and moved to and fro by some kind of mechanism; or a small perforated pipe was introduced into the can within a few inches from the bottom, and a current of cold air forced through the pipe, rising in bubbles through the water and emerging at the top, thereby producing a circulation in the can. All of these arrangements had the disadvantage that at the end of the operation of making the ice block, the bar, paddle, or pipe had to be removed to prevent being frozen into the ice, while otherwise the effect was good. Another proposition was to rock the can in the tank, thus agitating the water. None of these different arrangements, however, found favour in practical use. The moving gear for many hundreds, even thousands, of cans proved quite cumbersome. In removing the cans from the tanks this gear was in the way, and had likewise to be removed, and, on the whole, few and comparatively small plants have adopted any one system.

The plate system and the shallow stationary cells alone avoided the agitation of the water, and yet produced clear ice. But the freezing taking place only from one side, the process was so slow, and the plants became so large and expensive, that these systems also have found few users.

Another mode of making transparent ice is to deprive the water of its air before it goes into the cans. This can be done by long-continued boiling, or by exposing the water to a high vacuum, but better still by distillation under exclusion of the atmosphere. The result of this process has been found extremely satisfactory, and is to-day the one most in use. In order to economise in fuel, however, it has been found necessary to use the exhaust steam from the engine for the purpose of ice-making; and the steam, therefore, had to be deprived of the oil used in lubricating the steam cylinder. This has been effectually accomplished by steam filters of very simple construction. After condensation of the steam thus filtered, the condensed water is again filtered in order to entirely deodorise it. As a result, can ice produced in this manner is as good ice as can be made. It contains but a very thin stratum of porous ice in the centre, due to reabsorption of air in the can during freezing, but is better, purer, and more durable than any natural ice which can be bought. The ice is obtained in rectangular blocks of any desired size, and the waste by melting out of the moulds reduced to a minimum.

Ice-factory buildings may be made of wood or brick, depending upon the building regulations or fire restrictions of the city or locality, as well as upon the ambition and means of the owner. In most cases, a plain and a well-built wooden building is used, Figs. 52 and 53, which are the plan and elevation of a 50 ton factory, giving a good idea of the requirements. The building should be divided into the following compartments: boiler room, engine or ice-machine room, and tank room, ice-storage house, having an anteroom and a business office.

The condensing and distilling apparatus may be placed upon the roof, or in a special room; in either case they should be protected by a roof, with open sides of lattice or screen work through which air can circulate. A convenient platform for loading into waggons and weighing the ice is also required.

Special cold-storage rooms for preserving butter, eggs, and perishable products may be added to the premises

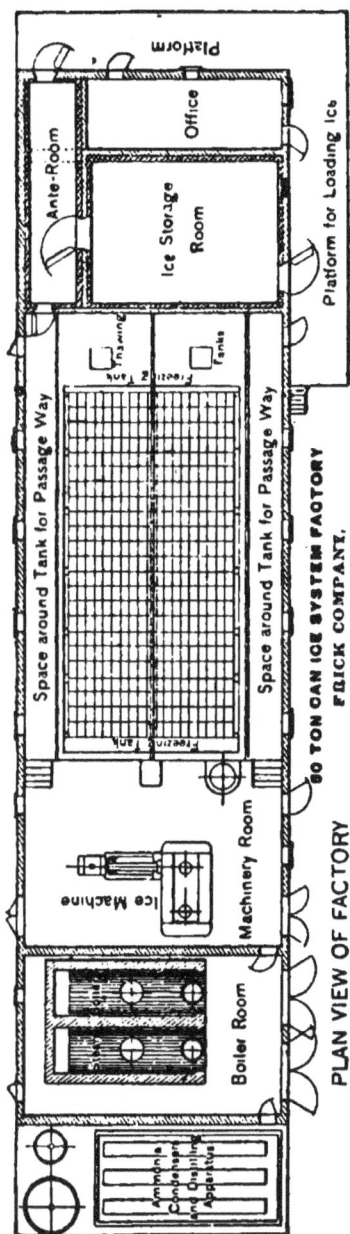

Fig. 52.

PLAN VIEW OF FACTORY

80 TON OAN ICE SYSTEM FACTORY
FRICK COMPANY,
ENGINEERS
WAYNESBORO FRANKLIN CO. PA.

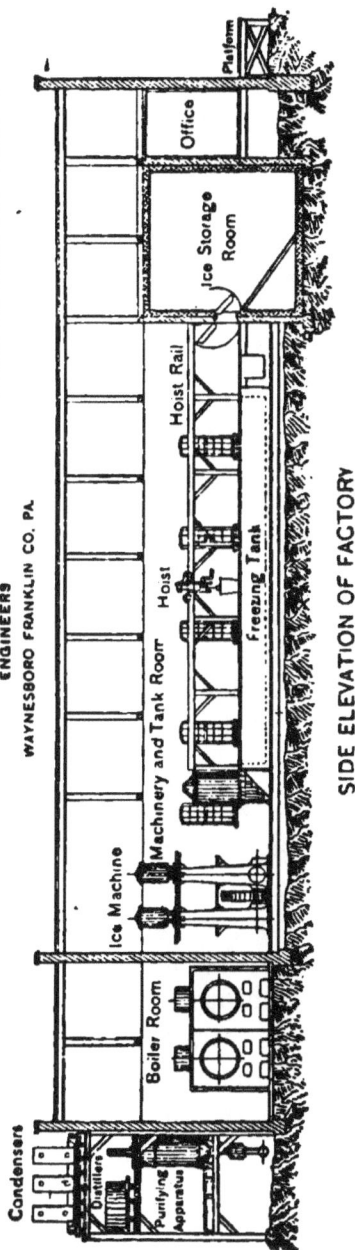

Fig. 53.

SIDE ELEVATION OF FACTORY

with profit, if the locality offers any inducement to provide for this kind of business.

A table showing the comparative cost of manufacturing each ton of ice under this system with machines the capacities of which range from one to one hundred tons per day, is to be found in the Appendix, Table E.

The arrangement of the De La Vergne ice-making plant is very clearly shown in Fig. 54. It will be readily understood that, as long as the principle of the system is not thereby disturbed, the various parts may be placed in relatively different positions to each other.

To begin at the compressor, which is shown to be a double-acting one and marked A. On the right-hand side the gas is drawn from the evaporating coils through the suction pipe B. By the action of the compressor the gas is discharged through the pipe C into the pressure tank D, where the oil, which we will follow later on, is dropped to the bottom. The upper half of this tank is provided with cast-iron baffle plates, which serve to more completely retain the oil and lodge it on the bottom. From the tank the gas, still hot by its compression, is sent through pipe E into the bottom pipe of the condenser F, where, by the action of cold water running over the pipes, the hot gas is first cooled, and then liquefied. The small liquid pipes G conduct the liquid ammonia through the liquid header H into the storage tank I, and from there it runs through the pipe J into the bottom of the separating tank K, which should be at all times at least three-quarters full. The small pipe L carries the liquid ammonia, in consequence of the pressure on it, to the expansion cock M, through which it is injected into the evaporating coils N, placed in the

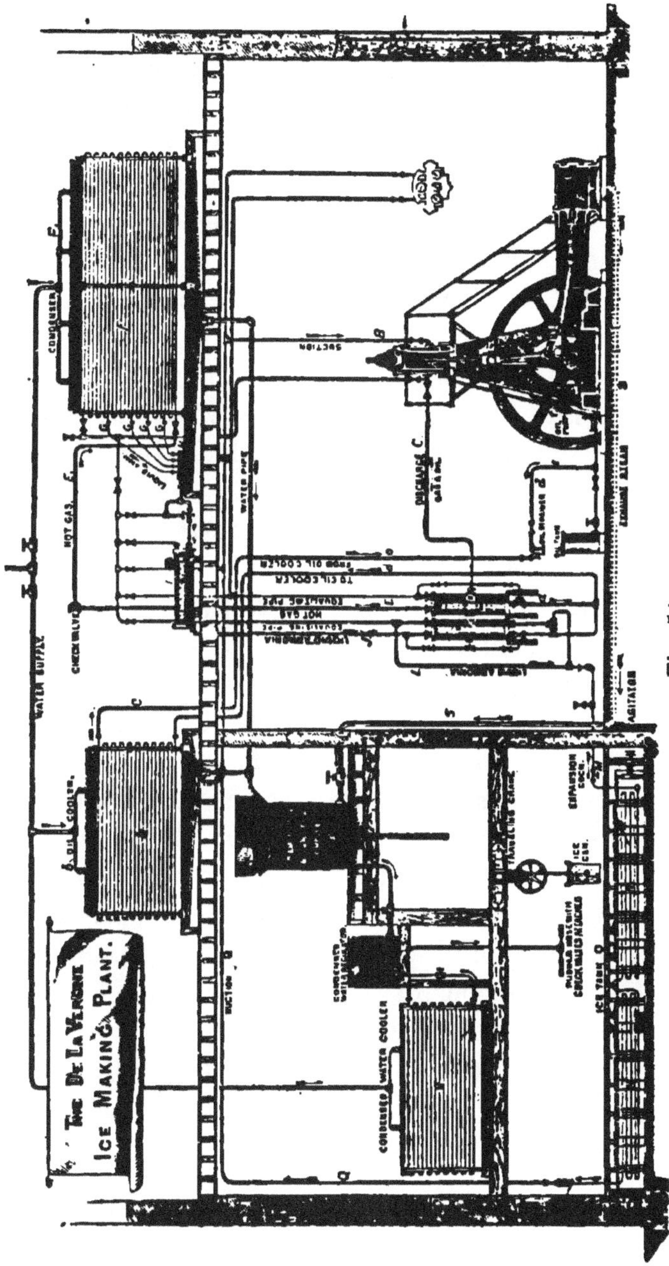
THE DE LA VERGNE ICE MAKING PLANT.
CONDENSER
WATER SUPPLY
CHECK VALVE
HOT GAS
SUCTION
DISCHARGE
WATER PIPE
EQUALIZING PIPE
CONDENSED WATER COOLER
ICE TANK
AGITATOR

Fig. 54.

freezing tank O. This tank contains a salt brine, non-congealable except at a temperature near zero; and, by the absorption of heat from this brine, the ammonia, in vaporising, cools it down to a temperature below 32 deg., say 17 deg. or 18 deg. Of the coils N there are a number side by side, leaving space enough between them to insert the galvanised iron ice cans P, which contain the water to be frozen. After evaporating in the coils N, and thereby having taken up heat from the brine, the ammonia gas now passes through the pipes Q and B back into the compressor from which we started. This is the entire cycle through which the ammonia passes.

It was found that the oil heated with the gas by compression was dropped into the bottom of tank D. From there it passes through the pipe a to the lowest pipe of the oil cooler b, similar in construction to the condenser, and, like it, cooled by cold water showered over it. After being cooled down in the oil cooler, it passes through pipe c, strainer d, and pipe e, into the oil pump f, which is so constructed that it distributes the cold oil into the compressor on either side of the piston during its compression stroke, *i.e.* in such a manner that no oil is furnished during the suction stroke of the piston, but only during the time of compressing, thereby cooling the gas during its period of heating. The hot oil, after leaving the compressor, now returns again, in company with the hot gas, to the tank d, and from there again enters on its course through the oil cooler, strainer, and oil pump, to the compressor.

It will be seen that both the ammonia and oil go through complete cycles, and that no waste of either is likely to occur except by leakage. In case, however, small traces of oil are carried along with the current of the gas from the pressure tank D into the condenser F, these small quantities flow along with the liquid ammonia into the separating tank K, where they collect at the bottom, the oil being heavier than liquid ammonia. When a certain amount of oil has collected here, it can be drawn off through the cock *g* and pipe *h*, and carried through the oil cooler back into the oil pump and compressor.

The steam from the steam cylinder marked R passes through the exhaust pipe S into the steam filter and condenser T, where it is purified and condensed. Out of the condenser T, it runs into the water-regulator tank U, from there through the condensed-water-cooling coil V, constructed like the ammonia condenser and oil cooler, and cooled by cold water, and is ultimately filled into the ice cans through rubber hose and cocks. After the cans have their contents frozen, the travelling crane transports them to the dip tank or sprinkler, where the block is melted out. The empty can is put back into its position in the freezing tank, refilled with water, and the process of making another block is commenced.

The arrangement of the fermenting room, Fig. 55, the beef chill room, Fig. 56, and the hog chill room, Fig. 57, as fitted by Messrs. Sterne & Co. differs very slightly in appearance from those of the Frick Co. The principal feature, and to which attention is directed, is the discs on the cooling pipes; these will be noticed on referring to the above illustrations.

Formerly Messrs. Sterne & Co. used pipes only to obtain the necessary cooling surface in the rooms to be

refrigerated; but since 1882 they have accomplished the
same object by means of cast-iron discs, which are made

Fig. 55.

in halves and attached to the expansion coils, after these
are all put up, by means of iron clips, which press the

two halves together against the pipes. The cooling
surface is thereby increased to such an extent that only

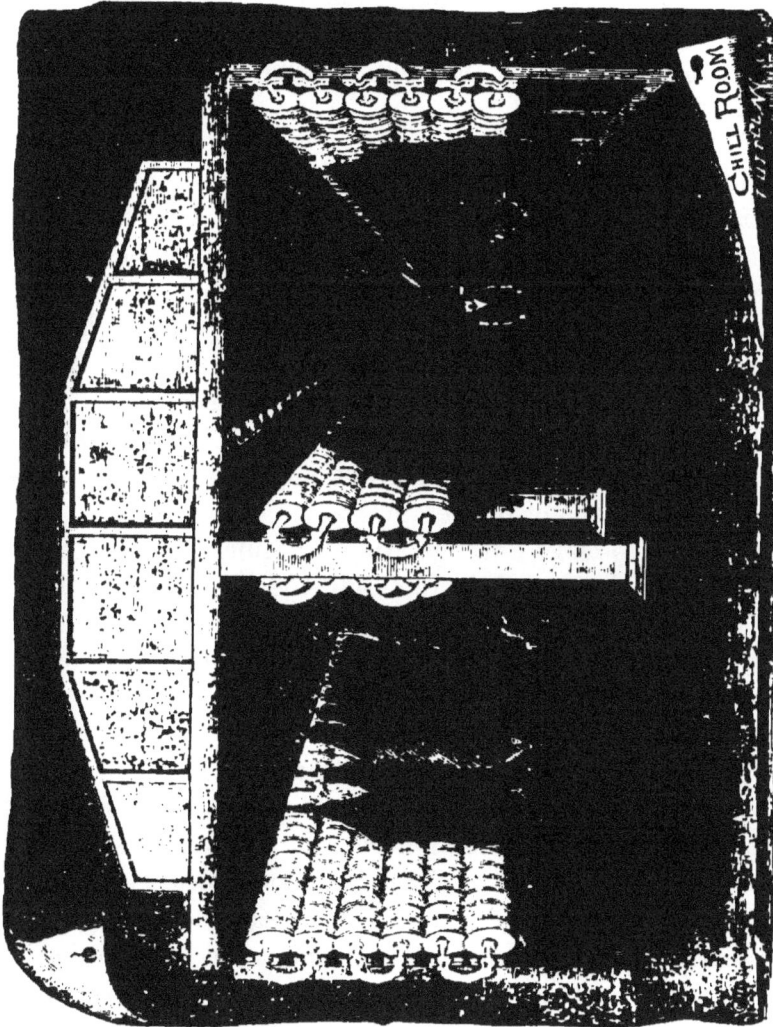

Fig. 56.

one foot of pipe is now required instead of four, thus
saving in room and first cost.

The application of the disc is based upon the principle now used in the most efficient of our modern steam radiators, in which the heating surface exposed to the air is increased by means of flanges and projections added to the outside surface of the radiator; thus exposing a larger heating surface than was attained with the old form of steam coils.

By applying these discs to steam coils, the same results could be obtained as with the modern steam radiator; the transmission of heat could be increased or diminished according to the number of discs applied to each lineal foot of pipe.

The results obtained are based upon the fact that heat is conducted with more rapidity by iron than by air. Whereas, one square inch of iron will transmit, say, 50 heat units per minute to another piece of iron attached to its surface, it will transmit but one heat unit, under similar conditions of temperature to air.

In order to make a refrigerating coil quick and effective in reducing the temperature of air, the air is brought in contact with as large a refrigerating surface as practice admits of, without, however, increasing the internal surface bathed with the chilled liquefied ammonia to more than is absolutely necessary.

Fig. 58 is a cross section of steamer engaged in the frozen meat trade and fitted with Messrs. J. & E. Hall's patent carbonic anhydride refrigerating machines.

For maintaining the insulated holds at 15 deg. Fahr. for frozen meat, or at 30 deg. for chilled meat, the brine-pipe system is considered the most advantageous. The brine cooled by the machine in the engine room is circulated by a small pump through wrought-iron pipes fixed on the

Its Principles and Management

under side of the deck over the hold or insulated space.
These pipes being usually placed between the deck beams,
thus occupy no valuable space, and are protected from
damage.   The brine pipes are divided into sections, each
section having a separate flow and return from the
machine room, where valves are placed for regulating the
quantity of cold brine in each section, as required by the

Fig. 58.

temperatures in the holds.   The pipes being only subject
to about 10 lbs. pressure, there is no risk of any leakage
from joints.

The presence of the cold brine pipes at the top of each
hold sets up a constant circulation of air, for as the air in
contact with the pipes becomes colder and consequently
heavier, it descends and is replaced by less cold air, which
is cooled in turn.

The pipes also have the effect of producing very dry cold air, any moisture in the air being condensed upon them.

The large quantity of cold brine in the holds acts as a storage of "cold," thus maintaining the holds at a low temperature for a considerable period after the machine is stopped. In most cases it is therefore necessary to run the machine for only a few hours daily. This advantage is obtained by no other system.

The machines do not require to be placed in proximity to the insulated holds or chambers, as the cold brine can be carried in insulated pipes for a considerable distance without appreciable loss. The machines are usually placed in the main engine-room. They are preferably of the duplex type, each half being practically independent and able by itself to maintain the necessary temperature. By this means a very large margin of safety is ensured.

There is no carbonic anhydride in the holds. It is contained only within the machine in the engine room, where, should a leak occur, it is not harmful or even disagreeable; in fact the whole contents of the machine may be allowed to escape into the engine room without any danger.

As carbonic anhydride, unlike ammonia, attacks no metals, the pipes in which it is condensed are made of copper to withstand the corroding action of the sea water.

Every part of the machine is tested to at least three times the working pressure.

An illustration of a meat store and freezing rooms is given in Fig. 59. These are constructed under Hall's patent wall system of meat freezing and chilling by radiation.

14

The brave cooled by the carbonic anhydride machine is circulated through hollow walls made of steel plates,

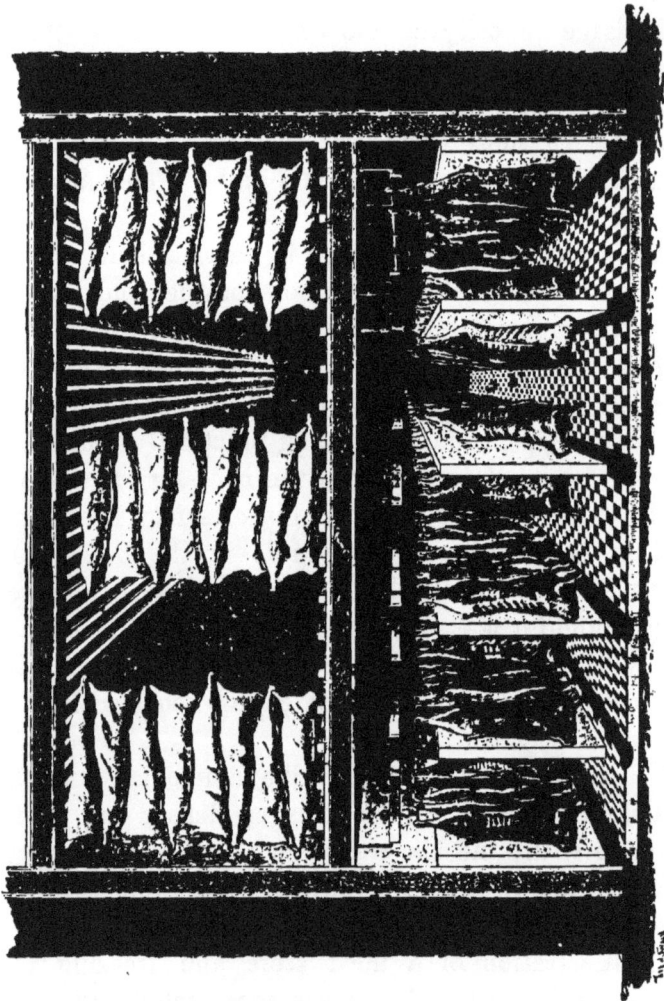

Fig. 59.

placed parallel to each other at short intervals.   In the passages between these walls the meat to be frozen or chilled is hung, and the very low temperature at which the walls are kept causes the heat to radiate from the meat, which becomes rapidly chilled or frozen.

For large works, J. & E. Hall Limited carry out a patent system by which the meat is hung on hooks connected together, forming a continuous chain, which is set in motion by gearing; in this way the meat requires no handling from the time that it is hung in the slaughter-house, till it arrives at the other end, chilled or frozen, as the case may be, and ready for despatching.

The sheep are placed 16 inches apart, the walls occupying less than 1 inch between the rows, so that the freezing rooms only occupy the usual space.

The rapidity of freezing and chilling by this patent method is found to be considerably greater than by any other, consequently the space set apart for freezing rooms is much reduced, and therefore the buildings are less costly. It, moreover, effects a great saving of labour, and depreciation of the meat by handling.

The stores containing the frozen sheep are maintained at the necessary temperature, either by the circulation of large volumes of cold dry air, cooled in the engine room, or, preferably, by means of brine cooled by the machine, and circulated in wrought-iron pipes placed under the ceiling of the cold chamber.

J. & E. Hall's carbonic-anhydride machines and their system have been found very effective for the importation of fruit, to which branch they have devoted special attention. The problem here is to keep the chambers at a constant and equable temperature, and to maintain a constant, though not too vigorous, circulation of air.

For making ice and preserving provisions on passenger

steamers, steam yachts, etc., Messrs. Hall supply a very
compact and serviceable installation.   The chambers,
Fig. 60, are constructed of an outer and
inner skin of two thicknesses, each of ¾
inch or 1 inch boards, tongued and
grooved, and put together air-tight, the
space between the two skins, about 6
inches or 8 inches, being filled with some
good non-conducting material, such as
charcoal.

Separate brine services are fitted for
producing the various temperatures required in the meat rooms, vegetable rooms,
butter rooms, fruit rooms, as well as for
the making of ice and for water cooling,
and are controlled from the refrigerating
machine, which may be placed in any
available corner in the main engine room,
where it comes under the eye of the
engineer on watch.   By this system the
chambers can be located in any convenient part of the ship; if necessary, at
a considerable distance from the engine
room.   The ice-making box and water
cooler are sometimes placed in the steward's
pantry.

These small machines are perfectly
simple, and can easily be worked by a
person of ordinary intelligence from the printed instructions supplied with each machine.   They need very little
attention, and the wear and tear and consequent need of
repairs is almost nil.

The complete charge of carbonic anhydride in the machine is so small that it might be allowed, under ordinary conditions, to escape into the engine room, without the slightest inconvenience. A patent safety valve is fitted so that no mistake or neglect on the part of the attendant can cause anything like an explosion.

Messrs. Hall's carbonic anhydride refrigerating machines are particularly applicable for use in breweries and distilleries, Fig. 61, as the refrigerating material used, carbonic acid, is given off in immense quantities from the vats during fermentation, and is thus a material with which brewers are well acquainted.

The cooler consists of a cylindrical tank, through which the water to be cooled is passed; within this tank are the coils containing the liquid carbonic acid supplied by the condenser of the machine; this liquid returns into the gaseous condition as fast as it can pick up heat from the water to be cooled, after which it is recompressed, and liquefied again in the condenser.

The water can be cooled from any temperature, 90 deg. or 100 deg. if necessary, to any temperature above 32 deg., its fall in temperature depending upon the quantity passed through the evaporator: thus, by regulating the outlet stop valve on the evaporator tank, the temperature of the water cooled can be varied to suit requirements.

When the machine is to be used for cooling fermenting tuns or storage cellars below 40 deg. Fahr., it is necessary to use brine instead of water, which is circulated, at, say, 25 deg. Fahr. through pipes placed under the ceiling of the space to be cooled. By this means the air surrounding the pipes is rapidly cooled; its gravity being thereby increased, it falls, displacing the warmer and lighter air

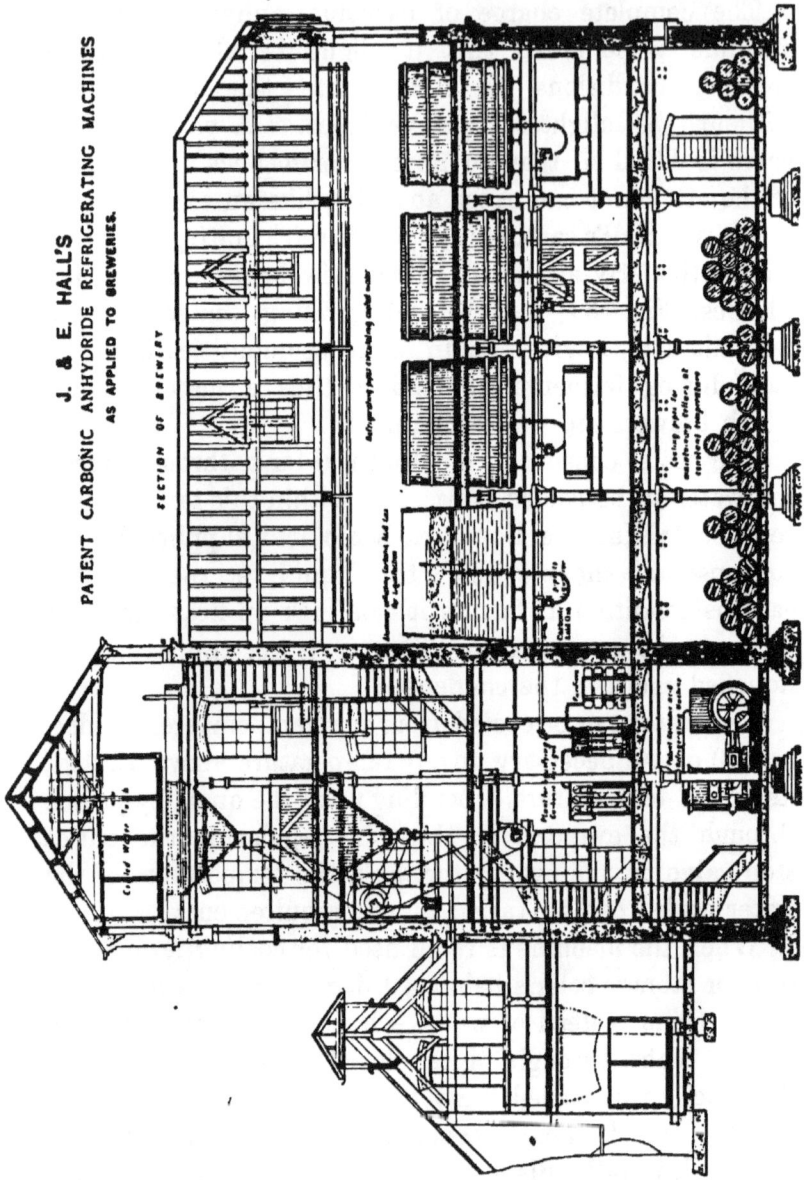

J. & E. HALL'S
PATENT CARBONIC ANHYDRIDE REFRIGERATING MACHINES
AS APPLIED TO BREWERIES.

SECTION OF BREWERY

Fig. 61.

which in turn is cooled, thus producing a thorough circulation.

Arrangements can be made for the collection and liquefaction of the carbonic acid given off by the tuns during fermentation, and the gas can be purified by a simple and inexpensive method, when it is said to compete most favourably with gas produced by other systems, in all

Fig. 62.

of which chemicals and a cumbrous and expensive plant are necessary.

The Pulsometer Engineering Company Limited also manufacture a considerable number of refrigerating machines for domestic use; for the preservation of fish and provisions; and for manufacturing purposes. Fig. 62 is an illustration of an engine and pump room containing

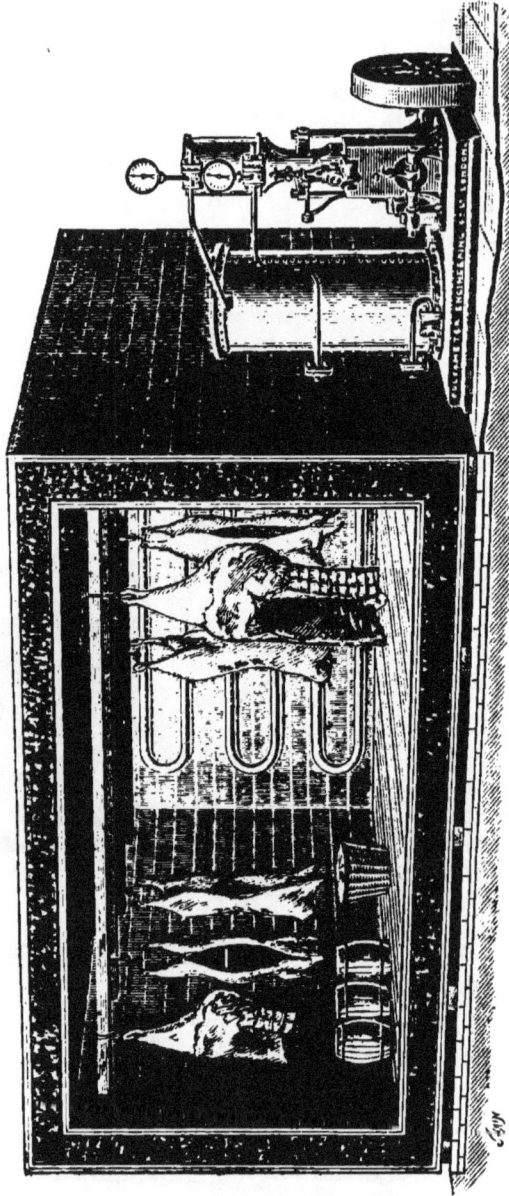

Fig. 63.

their regenerative ice plant, making 25 tons of ice per day. For very small installations, they supply vertical or horizontal high-pressure engines or other motors. For large installations, they provide compound-condensing engines, generally of the horizontal type, of extra size, as in the illustration, so that there may be an ample margin of power for hot weather, and in addition that the best results as to economy may be obtained from an early cut off. Where the highest economy is desirable the Pulsometer Co. advise the adoption of triple or quadruple-expansion engines. In all their engines provision is made for lubrication while running, in order to avoid the necessity of stopping for that purpose. The adjustment of all working parts is also provided for, thus preventing long stoppages for taking up wear.

Fig. 63 is an illustration of one of their cold stores specially designed for food preservation. These stores are of standard sizes, varying from 2000 to 70,000 cubic feet capacity, and capable of maintaining from 15 to 400 tons of frozen meat at 24 deg. Fahr. in England. The Pulsometer Co. claim that a machine on their regenerative system will maintain 60,000 frozen carcases, at about 20 deg. Fahr., with an expenditure of 2 tons of coal per twenty-four hours.

CHAPTER VII

It is important in ice-making, where the greatest production with the least possible cost is desired, to run the
plant day and night, making the operation continuous.
Not only this, but the conditions necessary to insure the
best quality and greatest quantity should be strictly
observed. For instance, in drawing the ice, it must
be done with regularity, so many cans each hour, day and
night, and the drawing equally distributed over the whole
area of the tank. To assist in this, a figure should be
branded upon each cover, and a little practice will soon
enable the tankmen to keep a record of their work. It
is hardly necessary to suggest the reason why the ice
should be drawn systematically, the machine run at
regular speed, steam pressure, water supply, boiler feed,

and all temperatures maintained, as near as possible, uniform.

Keep all parts of the apparatus clean and in good order. Let nothing become foul or dirty about the distilling apparatus, ice cans or tanks. Means are provided for purging the entire distilling system by steam, and the use of a scrubbing brush and harmless solvents is recommended. Change the charge in the filters as often as required.

THE ICE FACTORY CREW, divided into two watches, each with a 12-hour watch, on a 50-ton ice plant, for instance, consists of one engineer, one oiler, one fireman and coal passer, three tankmen, one ice-house man, and one general helper on each watch.

THE OPERATING EXPENSES OF A FACTORY are made up of cost of fuel, light, oil and waste, slight loss of chemicals, sundry small repairs, salary of superintendent and engineer, with wages for fireman, tankmen, and other labour. It is a paying investment to employ good men and the best fuel obtainable. For particulars of cost of the various items in running Eclipse machines, see Appendix, Table E.

The following are a few simple directions for testing and charging the Eclipse and other similar systems of ammonia refrigerating machinery :—

It is important, before introducing the charge of gas into the machine system, to carefully test every part of the apparatus, and make it thoroughly tight under at least 300 lbs. air pressure, which pressure may be obtained by working the ammonia compressor and allowing free air to flow into the suction side of the pump by opening special valves provided for this purpose, the entire system being thus filled with compressed air at the desired pressure. While this pressure is being maintained, a search is instituted for leaks, every pipe, joint, and square inch of surface being tediously scrutinised. One method is to cover all surfaces with a thick lather of soap, leaks showing themselves by formation of soap bubbles. In the case of condenser and brine-tank coils, the tanks are allowed to fill with water, the bubbles of air escaping through the water locating the leak. It is important that the apparatus be thoroughly tight, for while each separate piece is carefully tested at the works, transportation and handling may damage them; besides a few joints have to be made on the premises, and it is necessary to go over the entire surface to make certain. While the machine is engaged in pumping air into the system, advantage should always be taken of this opportunity to purge the system of all dirt and moisture. To do this properly, valves are provided so that the apparatus may be blown out by sections, removing valve bonnets and loosening joints for this purpose. By this means it is positively ascertained that each pipe, valve, and space is strictly clean and purged of all dirt and traces of moisture.

A final test may then be made by pumping a pressure of 300 lbs. upon the entire system, and allowing the apparatus to stand for some hours, estimating the leakage, if any, by noting the decrease of pressure as shown by the pressure gauge connected to the system. The air pressure will shrink somewhat at first, by reason of losing heat gained during compression by the pumps. As soon as the air parts with its heat and returns to its normal temperature, the gauge will come to a standstill and remain at a fixed point (depending upon the barometer and changing temperature of the room), if the system is tight.

Do not charge the system until it is well cleansed, purged, and tight.

PREPARING TO CHARGE.—After the machinery has been made perfectly tight, air must be exhausted from the entire system, by working the pumps and discharging the air through the valves provided for this purpose on the pump domes. When the escape of air ceases and the pressure gauges show a full vacuum, it is well to close all outlets and allow the machinery to stand for some time, to test the capacity of the apparatus to withstand external pressure without leakage ; in some cases it has been discovered that parts while tight from internal pressure, owing to loose particles lodging over the leaks and acting as plugs to prevent the leak showing, give way and disclose the leakage, when subjected to an external pressure.

INTRODUCING THE CHARGE.—To introduce the charge connect the flask of ammonia, as per directions given with each flask, to the charging valve, the gauge still showing a vacuum, close the expansion valve in main liquid pipe connecting the receiver to the brine tanks. Then open the valve on ammonia flask, and allow the liquid to be exhausted into the system. We recommend placing the flask on small platform scales, in order to weigh the contents and know positively when the flask is exhausted. The machine may be run all this time at a slow speed, with the discharge and suction hand-stop valves open wide. As one flask is exhausted, place another on the scales, and continue until the liquid receiver is shown to be partly full, by the glass gauge thereon. Then shut the charging valve, and open and regulate the main expansion valve; the machine is then sufficiently charged to do work, as shown by the pressure gauges and gradual cooling of the brine and frosting of expansion pipe leading to brine-tank coils.

While the system is being charged, water is allowed to flow on the condenser, and men are employed in searching further for leaks, which can readily be detected by the sense of smell, each point being again gone over.

Ammonia is a great solvent, and in some cases leaks may be opened up by reason of the gas-dissolving substances that may have stopped defective places and withstood the air test.

ACTION OF AMMONIA ON METALS.—Ammonia in itself is a slight lubricant, and has no effect whatever on the iron or steel, of which the machinery is constructed. It will eventually purge and scour the entire system clean to the metal surfaces, the loose foreign matter being caught in the separators and interceptors provided for this purpose.

AIR IN THE SYSTEM.—Carelessness in regulating the expansion valve, and needlessly pumping a vacuum on the brine tank, or carelessly allowing leaky stuffing-boxes, may allow air to get into the system, as will also taking the apparatus apart without expelling the air, before the reintroduction of the ammonia gas. The presence of air in considerable quantity is readily noticed by an expert, by the intermittent action of the expansion valve and singing noise, rise of condensing pressure, loss of efficiency in the condenser, etc. Purging valves are provided on the condenser and other points to allow the imprisoned air to escape, and restore the apparatus to its normal condition of pressure and efficiency.

PACKING THE PISTON RODS.—This is an important operation, and much depends upon the method employed. Some engineers have a habit of filling the stuffing-box with coils of packing cut so short that the rings leave a space and do not join at the ends, and they further persist in pushing the packing in with their fingers, and trusting to a vigorous screwing up on the gland to send them home. This is a poor way. Better put each ring of packing in separately and drive it home with a stout packing stick and wooden mallet, break joints with the rings and continue to pack and ram home until the box is full. This method calls for a very light pressure of the gland, and in many cases will wear a whole season.

It is a good sign when the main discharge pipes from the pump are hot to the touch, as it indicates efficient action of the compressors. This is due to the mechanical work of compression, manifesting itself in the form of heat, and, taken in connection with a regular action of the valves and steady working pressures, is a good indication.

TO REGULATE TEMPERATURES.—Increasing the speed of machine or increasing back pressure, or both, will effect this result, and *vice versâ*.

Increasing back pressure means more gas pumped per stroke, owing to the density of the gas increasing with the pressure. The back pressure is regulated somewhat by the required temperature in brine tank; there is no difficulty in running at 35 lbs. by gauge with the Eclipse brine tanks, owing to a peculiar distribution and large allowance of evaporating coils. It is best economy, however, to carry the brine no lower than necessary to secure proper temperatures in the rooms.   The regulation of back pressure is done by the expansion valve placed in the main liquid pipe, between the receiver and evaporating coils.

In making brine solution for brine tanks and for testing the density of the brine, the following instructions should be followed :—

Use medium-ground pure salt.   Buy in bags for convenience in handling.   Allow about 3 lbs. of salt for

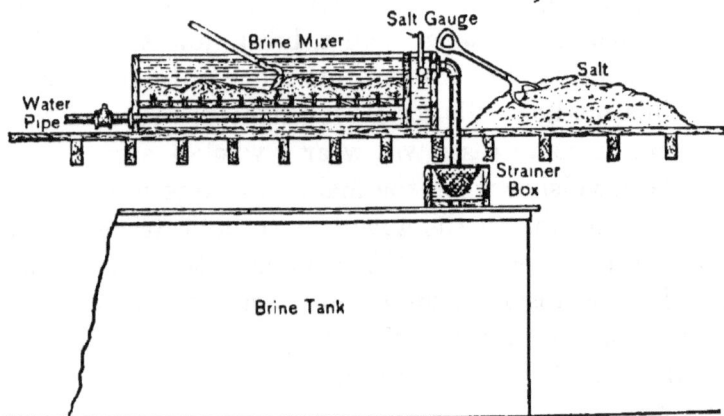

Fig. 64.

each gallon of water.   Continue to dissolve the salt in the brine-tank water until it reaches a density of 85 to 90 deg. by salt gauge.   If the brine is not of the full strength, low temperatures cannot be obtained, because the brine will freeze upon the evaporating coils, forming a thick coating of ice, which, to a large extent, acts as insulation.   Salt is simply added to the water in the brine tank to prevent its freezing, and to insure more intimate contact of the contents of brine tank with evaporating coils. The stronger the brine, the lower temperatures can be obtained without freezing.

In making the brine it is well to use a water-tight box, say 4 feet wide, 8 feet long, and 2 feet high, with a perforated false bottom and compartment at end.

Locate the brine mixer at a point above the brine tank, Fig. 64. Connect the space under the false bottom with your water supply, extending the pipe lengthways of the box and perforated at each side to ensure an equal distribution of water over the entire bottom surface ; use a valve in water supply pipe. Near the top of the brine mixer at end compartment, put in an overflow with large strainer to keep back the dirt and salt, and connect with this a pipe, say 3 inches diameter, with salt catcher at bottom leading into the brine tank. Use a hoe or a shovel to stir the contents. When all is ready, partly fill the box with water, dump the salt from the bags on the floor alongside and shovel into the brine mixer, or dump direct from bags into the brine mixer as fast as it will dissolve ; regulate the water supply to ensure the brine being always of the right strength as it runs into the brine tank : this point must be carefully noticed.

Filling the brine tank with water and attempting to dissolve the salt directly therein is not satisfactory, as quantities of salt settle on the tank bottom and coils, forming a hard cake.

It is a good plan, when it is desired to strengthen the brine, to suspend bags of salt in the tank, the salt dissolving from the bags as fast as required ; or the return brine from the pumps may be allowed to circulate through the brine maker, keeping the same supplied with salt.

Always see that the pump pistons of the Eclipse com-

15

pressors are adjusted so that they work close against the upper head. Do not forget this, as it is a great saving in power, and may make a great difference in the amount of work you can do.

The following instructions are for starting and working refrigerating and ice - making machines on the Linde system. These instructions would, however, with slight modifications, be applicable to other makes of machines working on the same principle :—

I. GENERAL DESCRIPTION.

The Linde system of refrigeration is based upon the evaporation of pure liquid ammonia, and the subsequent compression and liquefaction of the vapour thus formed.

The same charge of the chemical performs the circuit over and over again, the ammonia in one part of the machine being in the liquid state, and in another part in the state of vapour.

The work of compressing the ammonia vapour is done by a pump, which is driven by a steam engine, or in some other manner.

The liquefaction of the ammonia takes place within the ammonia-condenser coils. These are contained in a tank through which cooling water is continually passing while the machine is at work. The compressor forces the compressed vapour into a distributing piece connecting the tops of the coils, from whence it passes into the various coils. In these the ammonia is liquefied, and in that state it is collected and passed on to the regulating valve. It is important that all ammonia conveyed to the regulating valve is really liquid ammonia, and that it does not contain any bubbles or vapour. If vapour should be present, in consequence of there being an insufficient charge of ammonia, or from any other cause, a considerable falling off in the refrigerating power of the plant will result.

The evaporation of the liquid ammonia takes place in the coils of the refrigerator, which are placed in the ice-making tank, or in a separate vessel or chamber, according to the special purpose of the plant. The ammonia enters at the bottom of the coils, and in passing through them the bulk of the liquid is transformed into vapour, which, with a small quantity of liquid in suspension, is conducted from the top of the coils back to the compressor, in order to be recompressed and converted into the liquid form in the condenser.

It will be obvious that there should be nothing but pure ammonia in the interior of the machine, and if any air, water, or other impurity be present in an appreciable quantity, the cooling action of the machine will thereby be diminished.

Machines on the Linde system are applied for making ice, for cooling liquids, for cooling air, and for all freezing and refrigerating purposes, either by means of brine pipes, by pipes which are kept at a low temperature by the direct evaporation of ammonia inside, or by the Linde patent disc system. The various operations hereafter described will apply to every application, though in some cases the mode of working may be slightly modified in some details. Many of the Linde machines for marine use, as well as some of those for land purposes, are made with patent compound compressors, in which the compression of the vapour is accomplished in two stages

These compressors require the same treatment as those of the ordinary double-acting type. They are sometimes made with two single-acting cylinders of unequal areas, and are sometimes of the ordinary double-acting type. In the latter case only one end of the compressor is used for drawing the vapours from the main refrigerating coils, final compression being accomplished in the other end. Ammonia vapours from a refrigerator worked at a comparatively high temperature are sometimes admitted to the delivery pipe between the first and second compression. The compound system of working is patented.

II. STARTING THE MACHINE.

1. *Air Compression.*—Before a Linde machine is charged with ammonia, an air-compression test must take place for the purpose of ascertaining and proving the tightness of every part of the apparatus. To make this test the following instructions are to be observed.

The stop valve, situated in the suction conduit close to the compressor, is closed, and all the other stop valves connecting the various parts of the apparatus with one another, as well as the regulating valve, are opened.

The $\frac{3}{8}$-inch cock on the suction pipe close to the compressor should now be opened, or, in the absence of this cock, the cover of one of the suction valve boxes may be loosened, and the compressor slowly set in motion. Air will then be drawn in and pressed into the apparatus. When loosening the cover, a piece of hard wood should be inserted between the valve box and the cover, to prevent the valve box being blown out. The compressor must be worked at a slow speed, and it must be carefully lubricated with a moderate amount of oil to prevent the piston rings seizing. A large quantity of oil is to be avoided. The air pressure should be gradually increased to about 250 lbs. per square inch; but if, during this operation, the delivery pipe or the stuffing box should get hot, the machine must be stopped and the compression recommenced after everything has cooled down again.

When the pressure above mentioned has been obtained, the compressor should be stopped, the ⅜-inch cock closed, or the cover joint made tight. A thorough examination of all pipes and apparatus subjected to the air pressure should then take place. Particular attention must be given to the condenser, and specially to the joints at the ends of the coils and other parts.

The air compression must be kept up until every part and every joint is made perfectly tight. It is to be noted that some diminution of pressure will always take place after stopping the machine, due to the cooling and consequent contraction in volume of the air, which is delivered from the compressor in a more or less heated condition.

2. *Evacuation of Air.*—Before evacuating, the ⅜-inch joint at the bottom of the oil collector should be opened, in order to discharge all oil that may have passed through the compressor during the air-compression test.

The compressor may then be blown off, by opening the ⅜-inch cock on the delivery pipe.

To discharge the compressed air from the condenser, the regulating valve and the ¾-inch cock between the regulating valve and condenser are shut. The joint on the regulating valve side of the ¾-inch cock is then broken, and when the opening is quite clear the $\frac{1}{4}$-inch cock is opened full and the air discharged.

To discharge the refrigerator, all the small cocks in the pipes leading to the separate coils in the refrigerator or ice generator are closed. The blank flanges on the distributing pieces of the refrigerator or ice generator should then be removed, after which each refrigerator coil may be blown off one after the other, by opening the small cocks corresponding to each. By this process, not only water, which may have remained in the coils, may be removed, but also any obstructions can be discovered.

In order to produce a vacuum in the apparatus, the stop valve nearest to the compressor in the delivery pipe should be shut. All other cocks and valves must be opened, and lastly the $\frac{3}{8}$-inch cock on the delivery pipe. The compressor should then be brought into slow action, whereby the air will be expelled by each stroke through the open cock, accompanied by a gradually diminishing noise. When no more air is expelled, the machine should be stopped, and the $\frac{3}{8}$-inch cock at once closed.

We now proceed with the directions for charging the machine with ammonia and commencing work :—

3. *Charging with Ammonia.*—Pure anhydrous ammonia for this purpose can be obtained in strong iron or steel bottles.

In charging the machine, the vessel containing the anhydrous ammonia should be placed in a vertical position near the compressor, valve downwards, and a suitable connecting pipe carefully fixed from the cock or valve of this vessel to a $\frac{3}{8}$-inch cock on the suction conduit of the compressor. Then, by opening the valve on the bottle, the ammonia should be allowed gradually to pass into the machine. If hoar-frost should begin to form on the exterior of the vessel it should be warmed by a gas flame, or by a few pieces of red-hot iron, to ensure the contents being entirely discharged.

The vessels containing the anhydrous ammonia must always be stored in a cool place, and protected from injury. They must be moved with caution. Care must also be exercised in connecting the bottles to the machine, as the entire charge may be lost through a defect in the connection.

In cases where it is not convenient to use pure anhydrous ammonia, a separate distilling apparatus is supplied, whereby anhydrous ammonia is produced from the ordinary aqua-ammonia of commerce. The aqua-ammonia employed should be quite clear, and free from tar and other foreign matters. The specific gravity ought to be if possible ·880, and in any case not higher than ·900. The utilisation of weaker solutions is not advantageous.

The distiller is charged in two ways, according to the arrangement of the apparatus. The aqua-ammonia may be conveyed by means of a hand pump from a measuring vessel gauged to measure certain quantities, into the boiler of the distiller, after the measuring vessel has received the stipulated quantity from the carboy, or, more especially in the case of large machines, the hand pump may be used for putting the aqua-ammonia into the measuring vessel, and the liquor then passed over into the distiller by means of the ammonia pressure derived from the condenser, by opening the passage between the measuring vessel and the ammonia delivery conduit. For the first charge, however, this pressure is not in existence, and also not necessary, because the aqua-ammonia is passed into the distiller in consequence of the vacuum present in the latter.

The separate charges of aqua-ammonia are, for distiller No. VII., about eighteen gallons; No. VI., eleven gallons; No. V., eight gallons; No. IV., five and a half gallons, and for still smaller apparatus four gallons.

When the distiller has been charged, and all communication with the outside atmosphere has been shut off, the ammonia vapours are at once generated and allowed to pass into the evacuated machine, by opening the cock on the top of the apparatus, and a small amount of steam is sent through the heating coil, so as to slightly warm the steam exhaust pipe. Simultaneously the cooling water is turned on. The pressure gauges of the distiller and of the machine will soon show pressure exceeding that of the atmosphere. The distilling cock should then be entirely opened and more steam admitted, until the steam pressure is about 10 lbs. per square inch less than the ammonia pressure. If the distilling process is going on properly, the pipe leading from the distiller feels warm only for a short part of its length. If, however, too much steam be introduced, the generation of ammonia vapours takes place so rapidly that the liquor to be distilled is also carried over. Should this occur, the pipe of the distiller becomes hot throughout its whole length, the cooling water also runs away hot, and the distilling process must be at once stopped. After the distilling process has been going on for several hours, a sample of the liquid should be taken, and after being cooled down to 60 deg. Fahr., its density should be ascertained. The distilling process is finished when the specific gravity of the solution is ·970. When this point is reached, the distiller cock should be shut, and the distiller discharged by slightly opening the discharge cock. The exhausted liquor must then be measured both as to quantity and specific gravity, and the figures compared with those showing the quantity and specific gravity of the strong solution filled into the distiller. The quantity of ammonia which has actually entered the machine is then found by reference to a table. The amount of water carried over, if any, is also checked at the same time. All four observations, *i.e.* the quantity of strong and exhausted liquid, and the specific gravity of strong and exhausted liquor, must be entered into a log-book.

The distillation of the succeeding charges is performed in a similar manner. The distilling cock, however, should be opened as soon as the pressure in the distiller has nearly reached the pressure in the refrigerator.

After several charges have been filled into the machine, the pressure will have risen to 70 or 80 lbs. per square inch, and it is then necessary to bring all the ammonia into the condenser, in order that the refrigerator pressure may be reduced, under which circumstance the distilling operation is more easily performed. For this purpose the regulating valve is closed, all other valves of the exhaust and delivery conduits opened, and the machine is slowly set to work. When the pressure in the refrigerator has fallen to zero, the compressor is stopped, and a valve in the delivery conduit closed. The distillation is then continued as before, the charges of ammonia are repeated, and the ammonia is, from time to time, brought from the refrigerator into the condenser as just mentioned, until the machine has been sufficiently charged to enable it to commence work.

4. *Commencing Work.*—It must first be seen that all valves between the compresser, condenser, and refrigerator, excepting the regulating valve, are quite open; and, secondly, the flywheel must be turned round once or twice completely by hand to ascertain that no impediment of any kind exists to the free action of all parts. The machine is then started, the regulating valve opened from $\frac{1}{16}$th to a full turn (according to size of machine), and the temperature of the delivery pipe on the compresser carefully observed. Should this pipe become warm, then the regulating valve must be opened further, but if the pipe remains cool, the regulating valve should be gradually closed, until it is in the position at which the delivery pipe has about the same temperature as the cooling water leaving the condenser. If the delivery pipe should be too hot, although the regulating valve is quite open, then the machine is not charged sufficiently with ammonia. More ammonia must therefore be introduced.

If the delivery pipe remains cool and the machine is kept running, the regulating valve should be closed from time to time, as the temperature in the refrigerator falls. With a lower temperature in the refrigerator, the work done by the machine is reduced, therefore the quantity of liquid passing through the regulating valve must be reduced.

If any further signs of insufficiency of ammonia be observed, additional charges should be added whilst the machine is at work.

We have now to set forth directions for the working of this machine, and how to deal with interruptions in its action. The paragraph on the breaking of joints we have printed in italics in order to draw special attention to it :—

III. Working the Machine.

5. *Regulation.*—The regulation of the machine, that is, the adjustment of the quantity of ammonia circulating through it, is exclusively performed by means of the regulating valve, as already explained. The indications as to the amount of liquid to be passed through are given by the temperature of the delivery pipe, which should be about the same as the temperature of the cooling water passing from the condenser. If the delivery pipe is too warm the regulating valve should be further opened, and if the pipe is too cold the valve should be closed.

As the efficiency of the plant depends to a large extent upon proper regulation of the liquid ammonia, this operation should be conducted with great care. Whether in opening or closing, the regulating valve should be moved gradually.

6. *Signs for full and proper Efficiency.*—Besides the fact of the machine performing its proper refrigerating duty, there are a number of indications which permit of the action of the machine being checked and controlled :—

The delivery pipe is slightly warm (see Regulation). The temperature indicated by the pressure gauge of the refrigerator is from 10 deg. to 20 deg. Fahr. lower than the actual temperature of the brine or water being refrigerated.

The temperature indicated by the pressure gauge of the condenser is from 15 deg. to 20 deg. Fahr.

higher than that of the cooling water running from the condenser.

The play of the pressure and suction gauge pointers is regular, showing distinctly every stroke of the piston.

By placing the ear near the regulating valve the continuous and uninterrupted sound of the liquid ammonia rushing through the valve is distinctly audible.

The pipes conveying the ammonia to and from each refrigerator coil are equally covered with frost.

7. *Rectification.*—The small quantity of the lubricating oil which is conveyed by the piston rod into the interior of the machine is finally deposited in the oil collector, and unless removed this oil would in time pass into the coils and materially reduce the efficiency of the machine. Consequently a more or less continuous removal of the oil is necessary. The withdrawal of the oil from the oil collector takes place by means of a rotary cock, which is worked by means of a cord from a pulley on the main shaft, and by means of which the oil is delivered into a vessel named the rectifier. So long as there is oil in the collector, this cock discharges a small quantity into the rectifier at each revolution, and it should always be kept working, unless the pipe between the cock and the rectifier should become covered with frost. When frost forms on this pipe it is a sign that ammonia, and not oil, is passing, and the cock should then be thrown out of gear for a few hours and then started again.

The oil should be discharged from the rectifier once or twice daily. To facilitate this a small quantity of warm, not hot, water may be poured into the rectifier jacket, the cock on the pipe leading to the suction of the compressor being open. After a short time this cock should be shut, and the oil may then be run out from the rectifier into a bucket or other receptacle, by opening the cock at the bottom. When all the oil has been discharged, the cock should be shut, and that leading to the suction of the compressor should be opened.

Rectification must be performed more or less constantly, and the quantities of oil extracted, as well as those consumed at the stuffing box, should be carefully noted down, because it is only by a comparison between these two figures that it is possible to see how far the machine is free from oil.

If a large quantity of oil is entering the compressor, it is an indication that the packing of the stuffing box requires attention.

In machines of small size, the rotary cock and rectifier are dispensed with. In such cases the oil can be removed by breaking a joint at a suitable part of the apparatus, taking care to remove the ammonia beforehand.

Occasionally also machines are supplied in which the rectifier is connected directly to the bottom of the oil collector, without the rotary cock. In this case the cock in the connecting pipe should be opened at least once every twenty-four hours, the cock between the rectifier and the suction pipe being at the same time closed.

8. *Additional Charges.*—Through unavoidable leakages at the stuffing box, small quantities of ammonia are lost which must be periodically returned to the machine. It is not possible to say from the outset how often this is necessary, because the amount of loss is dependent upon the condition of the stuffing box, and therefore upon

the attention paid to it. The principal signs indicating scarcity of ammonia are indicated by the necessity of the regulating valve being more open than usual, and a deficiency in the cooling action of the machine.

9. *Attention given to Machine.*—If it is found necessary to break a joint or open part of the machine, or even to repack a cock, it is desirable that all the ammonia in this part of the machine should be pumped out and put in another part. If, however, this is impossible, the ammonia must be discharged into water or into the air before doing anything whatever to the connections.

Through the working of the machine, and specially through the contraction of the indiarubber, the joints will get slack in time. It is therefore absolutely necessary that the engineer in charge must, from time to time (say at least once every month), go over each joint and tighten up the bolts. Otherwise it may happen that leakage of ammonia will occur. This is especially important in hot climates.

10. *Maintenance of Compressor.*—The stuffing box of the compressor must be treated with particular attention, in order to obtain an entirely ammonia-tight gland, and to prevent interruptions caused by the piston rod heating, and the packing being destroyed.

In case the stuffing box has to be repacked, all the cocks close to compressor, including the $\frac{3}{8}$-inch cock on the stuffing box, must be shut tight. A small piece of pipe with rubber hose is then connected to each of the two $\frac{3}{8}$-inch cocks situated on the suction and delivery pipes of the compressor. Both cocks should then be opened slowly, and the ammonia contained in the compressor discharged into a bucket full of water. After the ammonia has been discharged, the gland should be loosened, and the old packing taken out. The new plaited cotton packing should be cut into suitable lengths and soaked in compressor oil before it is put into the stuffing box. After four, five, or six cotton rings have been put at the bottom of the stuffing box, the small iron lantern should be put in, and on the top of this should be put three or four more cotton rings, and finally one of the special rubber insertion rings as supplied by the Linde Co. The gland should then be put on again and the whole tightly screwed up. Care should be taken to ensure the lantern being placed opposite the hole for the ⅜-inch cock.

The gland cap should also be packed with one or two of the special rubber insertion rings, in order to keep the oil in the stuffing box.

The gland of the stuffing box must be always equally tightened up by the three bolts, and in no case should the piston rod be permitted to become warm.

The small oil pump should act well and constantly, and the reservoir of compressor oil from which the pump draws the oil to be circulated must always contain a sufficient quantity of the lubricant.

The oil which has been already used and recovered by rectification may be used over again, but it must always be filtered through a layer of sawdust about 10 inches high, or preferably through cotton waste, and it should be used only as a mixture of one part of new oil and one part of rectified oil.

The consumption of oil for the small machines is hardly appreciable. Machine No. IV. will use about half a pint in twenty-four hours; No. V. about one pint; No. VI. two or three pints of oil. If more oil passes into the machine, it indicates that the packing of the stuffing box is not in proper order, or the surface of the piston rod is uneven.

The time of duration for any packing depends chiefly upon the condition of the piston rod. If this rod is everywhere of the same diameter, and if its surface be well polished and smooth (this being the case whenever proper attention is paid to it) the packing will last from three to four months, or even longer. If proper attention is not given, it may be that the packing has to be renewed every three or four weeks.

The stuffing-box chamber is connected either with the compressor suction conduits direct, or with the pipe leading from the rotary cock to rectifier. The small cock close to the stuffing box must always remain open during the working of the machine.

If the machine is well kept and properly worked, the cylinder and piston rings will remain brilliantly polished, showing hardly any wear, and a change of piston rings is not necessary for years. The covers should, however, be removed from time to time, and an examination made of the interior.

The valve covers should be taken off occasionally, and the valves examined. In case the valves should be marked on their surfaces, which will happen to a greater or less extent after the machine has been run for a few weeks, owing to particles of scale, etc., which are loosened from the interior of the pipes and pass through into the compresser, they should be ground smooth with fine emery powder.

11. *Maintenance of Condenser and Refrigerator.*—The coils of the condenser accumulate more or less deposit on their surface, according to the purity of the cooling water,

and these should be cleaned periodically by means of special brushes which allow of their being introduced from above into the spaces between the separate coils. If possible, the condenser coils should be taken out of the casing every one or two years, and each coil thoroughly examined and cleaned. Steam should be driven through each coil so as to burn out any oil and dirt.

In case of corrosion, the places should be thoroughly cleaned and filled with solder and covered with an iron hoop, and after that the coil should be tested to a water pressure of say 500 lbs. per square inch, so as to be sure of its safety. After examination and cleaning, and before putting the coils back into the tank, they should be thoroughly cleaned and painted with anti-corrosive paint, or well tarred.

The refrigerators must always be kept so full of the liquid to be refrigerated that the coils are properly covered.

The salt which is brought into the refrigerator must not form any deposit, and the brine should not contain salts which crystallise at low temperatures.

Deposit may be easily prevented by dissolving the salt, not in the generator or refrigerator itself, but in special vessels, from which the brine is drawn after the deposit has completely settled. The brine must be sufficiently concentrated that it will not freeze at the lowest temperature occurring in the ordinary work of the machine. As a rule, + 5 deg. Fahr. is not exceeded, and this corresponds to a solution containing 20 per cent. of salt, but this strength must also be regulated according to the working conditions.

12. *Maintenance of Ice Generator.*—The ice generator

16

must always be filled with brine to the level of the angle irons on which the moulds are travelled.

The ice moulds must be carefully handled, and must not receive any bulges, because these would hinder the discharge of the ice blocks, and cause great waste in thawing.

If the travelling gear for the moulds does not act properly, the cause for this must be ascertained. No extraordinary force should be permitted, as this would probably be followed by a fracture of the toothed pinions, permanent twisting of the shaft, etc.

The frames of the moulds must be kept in order, and the small wheels should be frequently lubricated with mineral oil, which will not freeze.

The rails of the travelling crane must be kept free from oil, as otherwise the crane will have a tendency to mount on the rails. The ropes of hemp or manilla used for the travellers worked by power must always be stretched before being laid on, by fastening them securely at each end, and then suspending a load from the centre of the rope. In the case of cranes worked by shafting, the lubrication of the bearings and movable supports must be carefully attended to.

The above description refers specially to the ice generators with labour-saving appliances, but with slight modification is equally applicable to the smaller machines with ordinary ice tanks.

IV. INTERRUPTIONS IN THE ACTION OF THE MACHINES.

13. *Deficiency or Surplus of Ammonia.*—If the machine does not contain enough ammonia its efficiency sinks, with the following signs :—

The delivery pipes are hot in spite of the regulating valve being full open.

The pressures in condenser and refrigerator are lower than usual.

In placing the ear against the regulating valve a rattling sound may be heard, caused by the passage of ammoniacal vapours together with liquid ammonia.

Inside the compressor the piston rings may (in extreme cases) be heard grinding, owing to insufficient lubrication.

If the deficiency of the ammonia be not very great, these phenomena are less easily observed, and they may only appear in part.

The case of too much ammonia being in the machine is much rarer, and this is characterised by too high a pressure in the condenser, provided no other reasons for this (such as air) are present.

14. *Oil or Water inside the Machine.*—If the rectification be neglected, or the distillation is not carried out with proper caution, oil or water may pass into the ammonia conduits, a circumstance which is attended by a reduction of efficiency and a diminishing of the pressures, particularly of the refrigerator pressure. The remedy to be applied is constant rectification. If considerable quantities of liquid be present, shocks will be produced within the compressor.

15. *Air in the Machine.*—If any part of the machine is opened air will enter, and it is necessary that this air be got rid of before the particular part is again charged with ammonia. If this should be overlooked, or if the air is only partly discharged, it will be immediately

noticed by the action of the machine. The condenser pressure will rise much higher than before, and if the machine is stopped the pressure in the condenser will not return to the pressure corresponding to the temperature of the cooling water. The best way of getting rid of the air is as follows:—

The compressor should be stopped and the regulating valve closed, but all other valves kept open. A small piece of pipe with rubber hose should be connected to the ⅜-inch cock situated on the top of the cast-iron distributing piece on the top of condenser. After the engine has been standing for, say, one hour, this cock should be opened slowly and the air discharged into the cooling water in condenser. As long as air is discharged, air bubbles will be noticed rising to the surface of the water, but as soon as the air is discharged these bubbles will cease, and instead a sharp rattling noise will be heard, caused by the water absorbing the ammonia vapour. In this case the cock must be shut at once, and the engine may then be started. As it is impossible to get all the air out at one manipulation, it is necessary to repeat the above process until all air is discharged. If the arrangement of the plant permits of it, it is desirable to pass the cooling water through the condenser during the time that the compressor is stopped.

16. *Deposits on the Interior of Coils, and Obstructions in Ammonia Conduits.*—The smaller ammonia pipes may sometimes become completely closed up by pieces of stuffing box packing being drawn into the compressor, or by pieces of solder or dirt which have been left in the tubes. If the conduit for the liquid ammonia between condenser and refrigerator is obstructed, the

delivery pipes will become hot, and the refrigerator will fall. In this case the pipes must be taken apart for examination and blown out by steam. Before doing this, it is well to ascertain whether the valves shutting this pipe off from the condenser and refrigerator hold perfectly tight. If the pipe from the collecting piece at the bottom of the condenser is obstructed, then the condenser must be discharged for the purpose of examining this part. If the small pipe for liquid ammonia leading to one of the refrigerator coils becomes closed, this pipe will not show a covering of frost, and its corresponding return pipe is also not covered. If there is less frost on the exterior of one pipe than on another, this particular coil acts very slightly or not at all. In the first instance endeavours may be made to clear such a coil by conducting the entire stream of ammonia through it, by closing the small cocks in the pipes leading the ammonia to the other coils. If this is not attended with good results, the particular coil must be examined as soon as opportunity offers, and be thoroughly cleansed by blowing air or steam through it. If the refrigerator coils do not act properly, a decrease in the refrigerator pressure is observable.

The obstruction of small ammonia conduits is sometimes caused by carelessness in making the joints. This should be avoided, as far as possible, by care during erection.

Any deposits on the interior of the condenser coils will increase the condenser pressure, because the condenser coils do not transmit heat so perfectly as before.

17. *Irregularities in the Action of the Interior Parts of Compressor.*—Should the valves or the piston leak

(in which case the efficiency decreases) the refrigerator pressure will rise and the condenser pressure fall. The regulating valve will require to be closed more than usual, and the refrigerator can be evacuated only with great difficulties after the regulating valve is entirely shut. The piston and the valves must in this case be examined, the old piston rings repaired or replaced by new ones, and rough valve surfaces be ground smooth or new valves inserted. Signs similar to the above are present if a valve only acts partially, or does not work at all, in consequence of a fracture of a valve spring, valve guide, or of the valve itself. Simultaneously the sound of the valve action becomes irregular, or shocks and blows are audible in the compressor. Heavy blows are also caused if the nut on the piston has not been fixed properly and has worked loose. In all these cases the compressor must be immediately stopped, and the cause be found out and removed before setting to work.

18. *Breaking Joints.*—*When it is necessary to open any part of the machine, the ammonia should be transferred to another part and there retained. Such ammonia as cannot be thus transferred should be discharged through a short, strong, indiarubber tube, into a vessel filled and kept replenished with cold water, by which it will readily be absorbed. Care must always be exercised to ensure all ammonia pressure being removed before any joints, flanges, etc., are broken. When breaking pipe and other connections, the bolts should be loosened with great caution, so as to permit of the excess of ammonia pressure being gradually relieved, and the joint should not be fully slackened until it is certain that the ammonia has escaped.*

Regarding the management of machines working upon the dry-air principle, almost every make requires different treatment, but at the same time there are certain well-defined points to be remembered and difficulties to be avoided which are practically the same in all machines of this class. It should also be borne in mind that between machines working on different systems, such as air, ammonia, ether, carbonic acid, etc., there is a great similiarity in their working and management, and consequently our hints as to the care of ammonia machines will apply in great measure to other classes of machines.

As previously described, the air is drawn from the holds into the dryer or cooler; it circulates around the tubes and from thence it is taken to the compression cylinder; after being compressed it goes into the condenser, and afterwards, passing through the expansion cylinder, it is exhausted into the holds.

The most important thing is to make sure that the cold air is properly circulated, and that it enters every nook and cranny of the hold. This is usually accomplished by means of a pipe or trunk which is fixed along one side of the hold, and as near the top as possible. This trunk is fitted with a large number of small sliding doors; at the machine end these are only opened to a very slight extent, but the openings gradually increase until at the extreme end of the hold they are full open. The object of this is to ensure that as much as possible of the cold air will be circulated at the most distant part of the hold, which is naturally the warmest.

The return or exhaust-air pipe or trunk is placed at the opposite side of the hold from the other, and is also fitted with numerous doors which are opened in a similar manner, the widest open at the extreme end of the pipe, and decreasing gradually towards the machine in such a manner that the most of the air is drawn from the warmest part of the hold.

The temperatures of exhaust air, and sea-water discharge, and engine room should be carefully noted and entered on the log slate at the beginning and end of every watch by the engineer in charge. It would be well to hang a few thermometers at various places in the holds for the purpose of checking the deck-pipe thermometers.

It is necessary to frequently clean out the snow box, otherwise the passages would become choked, and, if allowed to accumulate, the slide-valve ports also. The trunk with its doors should be thoroughly cleared of all snow at least once every day, and the state of the holds and the condition of the cargo examined each day by one of the engineers, who should undertake this duty in strict rotation.

It so happens that the compressor valves sometimes get out of order. These valves are generally of the mushroom type, and are fitted with helical springs. When they set fast or fail to act, it will, as a rule, be found necessary to regrind them carefully in their seats; the springs at the same time should also be inspected, in order to find out whether they are bearing equally all round, the tendency being to wear more on one side than the other, this being due to the valves working horizontally.

While it is of importance to keep the valves in thorough working order, it is essential that the clearances in the compressor should not be too great, for if this be permitted, the compressed air left in the cylinder will cause an excessive back pressure, and prevent it from being properly filled with air on the return stroke.

The exhaust-slide-valve face of the expansion cylinder should, if anything, be kept slightly round, the extreme degree of cold having a tendency to contract the surface ; this ought always to be tightened up when the engine is going, as it can then be very.nicely adjusted ; if tightened up when the engine is stopped, the result will very probably be a broken eccentric strap.

The pistons, slide valves, and all the air valves should be tested periodically to discover whether any of them are leaking ; this may be done by maintaining the air pressure after the engines are stopped, and if there are any leakages, it will be at once noticed by opening the indicator cocks. Placing the hand at the back of the valve at the end of the stroke will also find out whether there is any leakage.

Previous to any cargo being shipped, the insulation in the holds should be carefully inspected, as the charcoal might be displaced, or possibly some of the planking might be damaged ; in the former case, the charcoal space should be properly filled up, and for this purpose it would be wise to always carry a supply on board ; in the latter case, the damaged boards should be repaired or replaced.

When the vessel has finished loading, the hatches should be carefully caulked with oakum, so as to render them air-tight. The practice now coming into favour is to fit the hatches with indiarubber joints, which make a more serviceable and efficient job than the oakum.

Before starting or stopping the engines, all the drain cocks should be opened, in order to drain out any water that may have accumulated; were this water allowed to remain it would be turned into snow, which in accumulating would block up the passages and consequently prevent the free. circulation of the air, in addition to the labour and trouble incurred in removing the snow.

The lubrication of the different parts of the machinery should be carefully attended to, particularly the slide valves and pistons; the compressor valves should also be lubricated regularly, and care taken to give them sufficient oil.

In most vessels where refrigerating machinery is employed, a complete spare engine is fitted on board, and in case of accident this engine can be at once started to work. It is hardly necessary to say that it is the bounden duty of the engineer in charge to have the spare engine in thorough working order, and ready to be set agoing at a moment's notice.

Where the refrigerating plant is a small one, or from any circumstances whatever a spare engine is not carried, the chief engineer should insist upon having a complete set of spare gear or duplicated parts of the refrigerating machinery, such as crank shaft, connecting rod, eccentric rod, piston rod, valve spindle, valves, springs, etc., and keep them where they would be readily got at.

There are numerous parts we have not touched upon in the management of air-refrigerating plant, for the simple reason that the hints regarding the care of ammonia machines will apply equally to those on the dry-air principle, where it is not specially mentioned otherwise.

In order that the working of the machinery may be checked from time to time, it is desirable to keep accurate logs showing the daily performance. Examples of such log sheets for ice-making, liquid, and air-cooling machines are given in the Appendix. These forms should, however, be taken merely as suggestions, and each one, when necessary, should be adapted to the special circumstances of each case.

In conclusion, we would strongly advise all engineers to thoroughly master all the details of the machinery committed to their charge, and to endeavour to understand the reason why for every part. Theory is no doubt an excellent thing to acquire, but it is of little use unless accompanied by a corresponding amount of practice.

APPENDIX

TABLE A.

Freezing Mixtures.

Composition by weight.		*Reduction of Temperature in degrees Fahr.*
Ammonium chloride	5 parts	
Potassium nitrate	5 ,,	From + 50° to + 10° = 40°
Water	16 ,,	
Ammonium nitrate	1 part	From + 50° to + 4° = 46°
Water	1 ,,	
Ammonium chloride	5 parts	
Potassium nitrate	5 ,,	
Sodium sulphate	8 ,,	From + 50° to + 4° = 46°
Water	16 ,,	
Sodium sulphate	5 parts	From + 50° to + 3° = 47°
Sulphuric acid diluted	4 ,,	
Sodium sulphate	8 parts	From + 50° to 0° = 50°
Hydrochloric acid	9 ,,	
Sodium nitrate	3 parts	From + 50° to − 3° = 53°
Nitric acid diluted	2 ,,	
Ammonium nitrate	1 part	
Sodium carbonate	1 ,,	From + 50° to − 7° = 57°
Water	1 ,,	
Sodium sulphate	6 parts	
Ammonium chloride	4 ,,	
Potassium nitrate	2 ,,	From + 50° to − 10° = 60°
Nitric acid diluted	4 ,,	
Sodium phosphate	9 parts	From + 50° to − 12° = 62°
Nitric acid diluted	4 ,,	

(*See continuation on next page*)

TABLE A (*continued*).

Freezing Mixtures.

Composition by weight.		Reduction of Temperature in degrees Fahr.
Sodium sulphate	6 parts ⎫	
Ammonium nitrate . . .	5 ,, ⎬ From + 50° to − 40° = 90°	
Nitric acid diluted . . .	4 ,, ⎭	
Snow or pounded ice . . .	2 parts ⎫	To − 5°
Sodium chloride	1 part ⎭	
Snow or pounded ice . . .	5 parts ⎫	
Sodium chloride	2 ,, ⎬	To − 12°
Ammonium chloride . . .	1 part ⎭	
Snow or pounded ice . . .	24 parts ⎫	
Sodium chloride . . .	10 ,, ⎬	To − 18°
Ammonium chloride . ..	5 ,, ⎪	
Potassium nitrate . .	5 ,, ⎭	
Snow or pounded ice . .	12 parts ⎫	
Sodium chloride . . .	5 ,, ⎬	To − 25°
Ammonium nitrate . .	5 ,, ⎭	
Snow	3 parts ⎫	From + 32° to − 23° = 55°
Sulphuric acid diluted . .	2 ,, ⎭	
Snow	8 parts ⎫	From + 32° to − 27° = 59°
Hydrochloric acid . .	5 ,, ⎭	
Snow	7 parts ⎫	From + 32° to − 30° = 62°
Nitric acid diluted . .	4 ,, ⎭	
Snow	4 parts ⎫	From + 32° to − 40° = 72°
Calcium chloride . . .	5 ,, ⎭	
Snow	2 parts ⎫	From + 32° to − 50° = 82°
Calcium chloride crystallised	3 ,, ⎭	
Snow	3 parts ⎫	From + 32° to − 51° = 83°
Potash	4 ,, ⎭	

TABLE B.

Evaporation of Liquids.

Liquid or Gas.		Water.	An-hydrous Ammonia.	Sul-phuric Ether.	Methylic Ether.	Sul-phur Dioxide	Pictet's Liquid.
Specific Gravity of Vapour, compared with Air = 1·000		0·622	0·59	2·24	1·61	2·24	—.
		Fahr.	Fahr.	Fahr.	Fahr.	Fahr.	Fahr.
Boiling Point at atm. pressure		212°	− 37·3°	96°	− 10·5°	14°	− 2·2°
Latent Heat of vaporisation at atm. pressure		966	600	165	473	182	—
See Fig. 1.	Fahr.	Lbs.	Lbs.	Lbs.	Lbs.	Lbs.	Lbs.
	− 40°	—	—	—	—	—	—
Absolute	− 20°	—	19·4	—	12·0	5·7	. 11·6
Vapour	0°	—	30·0	1·5	18·7	9·8	15·4
Tensions	+ 20°	—	47·7	2·6	28·1	16·9	22·0
in	+ 32°	0·089	61·5	3·6	36·0	22·7	27·0
lbs.	+ 40°	0·122	73·0	4·5	42·5	27·3	31·3
per	+ 60°	0·254	108·0	7·2	61·0	41·4	44·0
square	+ 80°	0·503	152·4	10·9	86·1	60·2	60·0
inch	100°	0·942	210·6	16·2	118·0	84·5	79·1
at	120°	1·685	283·7	23·5	—	117·5	99·7
different	140°	2·879	—	33·5	—	—	—
tempera-	160°	4·731	—	45·6	—	—	—
tures.	180°	7·511	—	62·0	—	—	—
	200°	11·526	—	81·8	—	—	—
	212°	14·7	—	96·0	—	—	—

TABLE C.

PRESSURE IN POUNDS PER SQUARE INCH OF AMMONIA GAS, CORRESPONDING TO TEMPERATURES FAHRENHEIT NAMED. SATURATED AMMONIA.

Temperature Degree F.T.	Temperature Absolute T.	Pressure, p. (From a vacuum.) Lbs. per sq. foot	Pressure, p. (From a vacuum.) Lbs. per sq. inch	Heat of Vaporisation, Thermal Units. he	External Heat, Thermal Units. $\frac{p.v.}{J}$	Internal Heat, Thermal Units. $p = e^{\frac{p.v.}{J}}$	Volume of Vapour per lb. cub. ft. v	Volume of Liquid per lb. cub. ft. v_2	Weight of a cub. ft. of vapour pounds $\frac{1}{v}$	Gauge Pressures per sq. inch.
− 40	420·66	1540·0	10·69	579·67	48·25	531·42	24·38	·0234	·0411	0
− 35	425·66	1773·6	12·31	576·69	48·35	528·34	21·21	·0236	·0471	0
− 30	430·66	2035·8	14·13	573·69	48·85	524·84	18·67	·0237	·0535	0
− 25	435·66	2329·5	16·17	570·68	49·16	521·52	16·42	·0238	·0609	1·47
− 20	440·66	2656·4	18·45	567·67	49·44	518·23	14·48	·0240	·0690	3·75
− 15	445·66	3022·5	20·99	564·64	49·74	514·90	12·81	·0242	·0775	6·29
− 10	450·66	3428·0	23·77	561·61	50·05	511·56	11·36	·0243	·0880	9·07
− 5	455·66	3968·0	27·57	558·56	50·44	508·12	9·89	·0244	·1011	12·87
+ 0	460·66	4373·5	30·37	555·50	50·66	504·12	9·14	·0246	·1094	15·67
+ 5	465·66	4920·5	34·17	552·43	50·84	501·59	8·04	·0247	·1243	19·47
+ 10	470·66	5522·2	38·55	549·35	51·13	498·22	7·20	·0249	·1381	23·85
+ 15	475·66	6182·4	42·93	546·26	51·33	494·93	6·46	·0250	·1547	28·23

+ 20	480·66	6905·3	47·95	543·15	51·65	491·50	5·82	·0252	·1721	33·25
+ 25	485·66	7605·2	53·43	540·03	51·81	488·22	5·24	·0253	·1908	38·73
+ 30	490·66	8556·4	59·41	536·92	52·02	484·90	4·73	·0254	·2111	44·71
+ 35	495·66	9493·9	65·93	533·78	52·22	481·56	4·28	·0256	·2336	51·23
+ 40	500·66	10512·	73·00	530·63	52·42	478·21	3·88	·0257	·2577	58·30
+ 45	505·66	11616·	80·66	527·47	52·62	474·77	3·53	·0260	·2832	65·96
+ 50	510·66	12811·	88·96	524·30	52·82	471·44	3·21	·02601	·3115	74·26
+ 55	515·66	14102·	97·63	521·12	53·01	468·01	2·93	·02603	·3412	82·93
+ 60	520·66	15494·	107·60	517·93	53·21	464·76	2·67	·0265	·3745	92·90
+ 65	525·66	16994·	118·03	515·33	53·40	461·82	2·45	·0266	·4081	103·33
+ 70	530·66	18606·	129·21	511·52	53·67	457·95	2·24	·0268	·4664	114·51
+ 75	535·66	20839·	141·25	508·29	53·76	454·70	2·05	·0270	·4978	126·55
+ 80	540·66	22192·	154·11	504·66	53·96	450·75	1·89	·0272	·5291	139·41
+ 85	545·66	24172·	167·86	501·81	54·15	447·75	1·74	·0273	·5747	153·16
+ 90	550·66	26295·	182·80	498·11	54·28	443·70	1·61	·0274	·6211	168·10
+ 95	555·66	28566·	198·37	495·29	54·41	440·95	1·48	·0277	·6756	183·67
+100	560·66	30980·	215·14	491·50	54·54	437·35	1·36	·0279	·7353	200·44

TABLE D.

Table giving weights of aqueous vapour held in suspension by 100 lbs. of pure dry air when saturated, at different temperatures, and under the ordinary atmospheric pressure of 29·9 inches of mercury. (Partly abstracted from "A Practical Treatise on Heat," by T. Box; partly calculated by T. B. Lightfoot.)

See Fig. 30.

Temperature.	Weight of Vapour.	Temperature.	Weight of Vapour.
Fahr.	Lbs.	Fahr.	Lbs.
− 20°	0·0350	102°	4·547
− 10°	0·0574	112°	6·253
0°	0·0918	122°	8·584
10°	0·1418	132°	11·771
20°	0·2265	142°	16·170
32°	0·379	152°	22·465
42°	0·561	162°	31·713
52°	0·819	172°	46·338
62°	1·179	182°	71·300
72°	1·680	192°	122·643
82°	2·361	202°	280·230
92°	3·289	212°	Infinite.

N.B.—The weight in lbs. of the vapour mixed with 100 lbs. of pure air at any given temperature and pressure is given by the formula

$$\frac{62\cdot3\ E}{29\cdot9 - E} \times \frac{29\cdot9}{p}$$

where E = elastic force of the vapour at the given temperature, in inches of mercury (to be taken from Tables)
p = absolute pressure in inches of mercury
= 29·9 for ordinary atmospheric pressure.

TABLE E.—ICE MANUFACTURE.

APPROXIMATE COST OF OPERATING ICE FACTORIES.

Tons Ice per Day	Engineers $1·50 to $5·00 per Day		Night Eng'r or Oilers $1·50 per Day		Firemen $1·50 per Day		Tankmen and Labourers $1·00 per Day		Pipe Fitter or Machinist $2·50 per Day		Coal 15 Cts. per Cwt., or $3·00 per Ton		Oil Waste, Lights and Sundries	Daily Operating Expenses	Ice per Ton
	No.	$	No.	$	No.	$	No.	$	No.	$	Amount	$	$	$	$
1	1	$1·50	—	—	—	—	1	$1·00	—	—	900	$1·35	$0·50	$4·35	$4·35
2	1	1·50	—	—	—	—	1	1·00	—	—	1,500	2·25	·50	5·25	2·63
3	1	2·00	—	—	—	—	1	1·00	—	—	1,800	2·70	·50	6·20	2·10
4	1	1·75	1	$1·50	—	—	2	2·00	—	—	2,200	3·30	·75	8·30	2·08
5	1	2·00	1	1·50	—	—	2	2·00	—	—	2,500	3·75	1·00	10·25	2·05
6	1	2·00	1	1·50	—	—	2	2·00	—	—	2,700	4·05	1·00	10·55	1·76
7½	1	2·00	1	1·50	1	$1·50	2	2·00	—	—	3,200	4·80	1·25	13·05	1·74
10	1	2·50	1	1·50	2	3·00	2	2·00	—	—	3,600	5·40	1·25	15·65	1·57
12½	1	2·50	1	1·50	2	3·00	3	3·00	—	—	4,500	6·75	1·25	17·00	1·37
15	1	2·50	1	1·50	2	3·00	3	3·00	—	—	5,000	7·50	1·50	19·00	1·27
18	1	2·75	1	1·50	2	3·00	3	3·00	—	—	5,500	8·25	1·80	20·30	1·15
20	2	4·50	2	3·00	2	3·00	4	4·00	—	—	6,000	9·00	2·00	23·00	1·15
25	2	5·00	2	3·00	2	3·00	4	4·00	—	—	7,500	11·25	2·50	27·25	1·09
30	2	5·00	2	3·00	2	3·00	5	5·00	—	—	9,000	13·50	3·00	31·50	1·05
35	2	6·00	2	3·00	2	3·00	5	5·00	—	—	10,500	15·75	3·50	36·25	1·03
40	2	6·00	2	3·00	3	4·50	6	6·00	—	—	12,000	18·00	4·00	39·00	1·00
45	2	6·50	2	3·00	4	6·00	6	6·00	—	—	13,500	20·25	4·50	42·75	1·00
50	2	6·50	2	3·00	5	7·50	7	7·00	—	—	15,000	22·50	5·00	50·00	1·00
55	2	7·00	2	3·00	5	7·50	7	7·00	1	$2·50	16,500	24·75	5·50	53·75	1·00
60	2	7·50	2	3·00	5	7·50	8	8·00	1	2·50	18,000	27·00	6·00	58·50	1·00
65	2	7·50	2	3·00	5	7·50	8	8·00	1	2·50	19,500	29·25	6·50	64·25	1·00
70	2	8·00	2	3·00	6	9·00	9	9·00	1	2·50	21,000	31·50	7·00	67·00	·99
75	2	9·00	2	3·00	7	10·50	10	10·00	1	2·50	22,500	33·75	7·50	74·75	·98
80	3	10·00	2	3·00	—	—	10	10·00	2	5·00	24,000	36·00	8·00	78·50	·96
90	3	10·00	2	3·00	—	—	11	11·00	2	5·00	27,000	40·50	9·00	87·50	·95
100	3	10·00	2	3·00	—	—	12	12·00	2	5·00	30,000	45·00	10·00	95·50	·95

The rate of wages and cost of fuel in the locality where machinery is to be used will change above figures slightly. The total allowances named are large, which fact will be recognised by experts.

TABLE F.

TABLE FOR ADIABATIC COMPRESSION OR EXPANSION.

ABSOLUTE PRESSURE.		ABSOLUTE TEMPERATURE.		VOLUME.	
Ratio of greater to less. *(Expansion)*	Ratio of less to greater. *(Compression)*	Ratio of greater to less. *(Expansion)*	Ratio of less to greater. *(Compression)*	Ratio of greater to less. *(Compression)*	Ratio of less to greater. *(Expansion)*
1·2	·833	1·054	·948	1·138	·879
1·4	·714	1·102	·907	1·270	·788
1·6	·625	1·146	·873	1·396	·716
1·8	·556	1·186	·843	1·518	·659
2·0	·500	1·222	·818	1·636	·611
2·2	·454	1·257	·796	1·750	·571
2·4	·417	1·289	·776	1·862	·537
2·6	·385	1·319	·758	1·971	·507
2·8	·357	1·348	·742	2·077	·481
3·0	·333	1·375	·727	2·182	·458
3·2	·312	1·401	·714	2·284	·438
3·4	·294	1·426	·701	2·384	·419
3·6	·278	1·450	·690	2·483	·403
3·8	·263	1·473	·679	2·580	·388
4·0	·250	1·495	·669	2·676	·374
4·2	·238	1·516	·660	2·770	·361
4·4	·227	1·537	·651	2·863	·349
4·6	·217	1·557	·642	2·955	·338
4·8	·208	1·576	·635	3·046	·328
5·0	·200	1·595	·627	3·135	·319
6·0	·167	1·681	·595	3·569	·280
7·0	·143	1·758	·569	3·981	·251
8·0	·125	1·828	·547	4·377	·228
9·0	·111	1·891	·529	4·759	·210
10·0	·100	1·950	·513	5·129	·195

TABLE G.

TABLE OF PROPERTIES OF SATURATED STEAM.

Pressure per square inch by gauge.	Temperature in Fahrenheit degrees.	Latent heat in heat units.	Total heat in heat units from water at 32° F.	Relative volume or cubit feet of steam from one cubit foot of water.
5	227·917	954·415	1151·454	1220·3
10	240·000	945·325	1155·139	984·8
15	250·245	938·925	1158·263	826·8
20	259·176	932·152	1160·987	713·4
25	267·120	926·472	1163·410	628·2
30	274·296	921·334	1165·600	561·8
35	280·854	916·631	1167·600	508·5
40	286·897	912·290	1169·442	464·7
45	292·520	908·247	1171·158	428·5
50	297·777	904·462	1172·762	397·7
55	302·718	900·899	1174·269	371·2
60	307·388	897·526	1175·692	348·3
65	311·812	894·330	1177·042	328·3
70	316·021	891·286	1178·326	310·5
75	320·039	888·375	1179·551	294·7
80	323·884	885·588	1180·724	280·6
85	327·571	883·914	1181·849	267·9
90	331·113	880·342	1182·929	265·5
95	334·523	877·865	1183·970	246·0
100	337·814	875·472	1184·974	236·3
105	340·995	873·155	1185·944	227·6
110	344·074	870·911	1186·883	219·7
115	347·059	868·735	1187·794	212·3
125	352·757	864·566	1189·535	199·0
135	358·161	860·621	1191·180	187·5
145	363·277	856·874	1192·741	177·3
155	368·158	853·294	1194·228	168·4
165	372·822	· 849·869	1195·650	160·4
175	377·291	846·584	1197·013	153·4
185	381·573	843·432	1198·319	147·1
235	401·072	831·222	1203·735	114
285	418·225	819·610	1208·737	96
335	431·956	810·690	1212·580	83
385	444·919	800·198	1217·094	73

NOTE.—By the term Saturated Steam is meant steam as it is formed in contact with water.

TABLE H.

COMPARISONS OF THERMOMETER SCALES.

R. Reaumur or German scale.   + Means above zero on each scale.
C. Centigrade or French scale.   − Means below zero on each scale.
F. Fahrenheit or English scale.

R.	C.	F	R.	C.	F.
+80	+100	+212	+53	+66·25	+151·25
79	98·75	209·75	52	65	149
78	97·50	207·50	51	63·75	146·75
77	96·25	205·25	50	62·50	144·50
76	95	203	49	61·25	142·25
75	93·75	200·75	48	60	140
74	92·60	198·50	47	58·75	137·75
73	91·25	196·25	46	57·50	135·50
72	90	194	45	56·25	133·25
71	88·75	191·75	44	55	131
70	87·50	189·50	43	53·75	128·75
69	86·25	187·25	42	52·50	126·50
68	85	185	41	51·25	124·25
67	83·75	182·75	40	50	122
66	82·50	180·50	39	48·75	119·75
65	81·25	178·25	38	47·50	117·50
64	80	176	37	46·25	115·25
63	78·75	173·75	36	45	113
62	77·50	171·50	35	43·75	110·75
61	76·25	169·25	34	42·50	108·50
60	75	167	33	41·25	106·25
59	73·75	164·75	32	40	104
58	72·50	162·50	31	38·75	101·75
57	71·25	160·25	30	37·50	99·50
56	70	158	29	36·25	97·25
55	68·75	155·75	28	35	95
54	67·50	153·50	27	33·75	92·75

TABLE H (*continued*).

R.	C.	F.	R.	C.	F.
+26	+32·50	+90·50	−4	−5	+23
25	31·25	88·25	5	6·25	20·75
24	30	86	6	7·50	18·50
23	28·75	83·75	7	8·75	16·25
22	27·50	81·50	8	10	14
21	26·25	79·25	9	11·25	11·75
20	25	77	10	12·50	9·50
19	23·75	74·75	11	13·75	7·25
18	22·50	72·50	12	15	5
17	21·25	70·25	13	16·25	2·75
16	20	68	14	17·50	0·50
15	18·75	65·75	15	18·75	−1·75
14	17·50	63·50	16	20	4
13	16·25	61·25	17	21·25	6·25
12	15	59	18	22·50	8·50
11	13·75	56·75	19	23·75	10·75
10	12·50	54·50	20	25	13
9	11·25	52·25	21	26·25	15·25
8	10	50	22	27·50	17·50
7	8·75	47·75	23	28·75	19·75
6	7·50	45·50	24	30	22
5	6·25	43·25	25	31·25	24·25
4	5	41	26	32·50	26·50
3	3·75	38·75	27	33·75	28·75
2	2·50	36·50	28	35	31
1	1·25	34·25	29	38·25	33·25
0	0	32	30	35·50	35·50
−1	−1·25	29·75	31	38·75	37·75
2	2·50	27·50	32	40	40
3	3·75	25·25			

LOG SHEET FOR AN ICE-MAKING MACHINE.

Machine No._____ Sheet No._____ Date._____

Date.	Time.	Revol. of Compressor per minute.	Condenser Gauge.		Refrigerator Gauge.		Cooling Water.			Heat units in Condenser per hour.	Brine.		Weight of Ice drawn.	Gallons of Compressor Oil.		Remarks.
			Press.	Temp.	Press.	Temp.	Temp. in.	Temp. out.	Galls. per hour.		Temp. in.	Temp. out.		In	Out.	

LOG SHEET FOR A LIQUID-COOLING MACHINE.

Machine No._____ Sheet No._____ Date._____

Date.	Time.	Revol. of Compressor per minute.	Condenser Gauge.		Refrigerator Gauge.		Cooling Water.			Heat units in Condenser per hour.	Brine or Water.		Heat units in Refrigerator per hour.	Gallons of Compressor Oil.		Remarks.
			Press.	Temp.	Press.	Temp.	Temp. in.	Temp. out.	Galls. per hour.		Temp. in.	Temp. out.		In	Out	

LOG SHEET FOR AN AIR-COOLING MACHINE.

Machine No._____   Sheet No._____   Date._____

Date.	Time.	Revol. of Com- pressor per minute.	Condenser Gauge.		Refrigerator Gauge.		Cooling Water.			Heat units in Con- denser per hour.	Cub. ft. of air circu- lated per hour.	Air Temperatures.								Gallons of Com- pressor Oil.		Remarks.
			Press.	Temp.	Press.	Temp.	Temp. in.	Temp. out.	Galls. per hour.											In	Out	

SOME OF THE USERS OF

J. & E. HALL'S PATENT CARBONIC ANHYDRIDE REFRIGERATING MACHINES.

H.M. War Department.
H.M. Ordnance Department.
Victorian Government (Railways Department).
South Australian Government.
His Highness the Maharajah of Nepaul.
London & Indian Docks Joint Committee.
The Bolton Corporation.
Bombay Ice Manufacturing Company.
Ismay, Imrie, & Co., White Star Line—9 Installations.
Peninsular & Oriental Steam Navigation Company—
　　2 Installations.
Union Steamship Company Limited—13 Installations.
Hamburg American Packet Company—9 Installations.
Eastern Telegraph Co.—2 Installations.
Eastern & South African Telegraph Company.
Elder, Dempster, & Co.
Huddart, Parker, & Co.
Lamport & Holt.
Commercial Cable Company.
Alfred Holt.
Houlder Brothers, London.

J. & E. HALL LIMITED,

DARTFORD IRON WORKS, KENT, & 23 ST. SWITHIN'S LANE, LONDON, E.C.

Telegraphic Address: "Hallford, London."　　Telephone No. 1846.

ADVERTISEMENTS.

SOME OF THE SHIPS FITTED WITH

J. & E. HALL'S PATENT CARBONIC ANHYDRIDE REFRIGERATING MACHINES.

WHITE STAR LINE.

S.S. "Gothic."
S.S. "Ionic."
S.S. "Cevic."
S.S. "Cufic."
S.S. "Germanic."
S.S. "Brittanic."
S.S. "Bovic."
S.S. "Coptic."

UNION STEAMSHIP COMPANY.

S.S. "Norman."
S.S. "Guelph."
S.S. "Scot."
S.S. "Gaul."
S.S. "Greek."
S.S. "Goth."
S.S. "Trojan."
S.S. "Mexican."
S.S. "Tartar."
S.S. "Athenian."
S.S. "Spartan."
S.S. "Pretoria."
S.S. "Moor."

HOULDER BROTHERS & CO.

S.S. "Ovingdean Grange."

BIBBY BROTHERS & CO.

S.S. "Staffordshire."

P. & O. S. N. COMPANY.

S.S. "Simla."
S.S. "Nubia."

HAMBURG AMERICAN STEAM PACKET COMPANY.

S.S. "Furst Bismarck."
S.S. "Persia."
S.S. "Prussia."
S.S. "Normannia."
S.S. "Columbia."
S.S. "Augusta Victoria."
S.S. "Patria."
S.S. "Phœnicia."
S.S. Now Building.

NELSON LINE OF RIVER PLATE STEAMERS.

S.S. "Highland Chief."
S.S. "Highland Mary.
S.S. "Highland Lassie."
S.S. "Highland Glen."

EASTERN TELEGRAPH CO.

S.S. "Duplex."
S.S. "Chiltern."

OVER 50 SHIPS FITTED.

J. & E. HALL LIMITED,

DARTFORD IRONWORKS, KENT, & 23 ST. SWITHIN'S LANE, LONDON, E.C.

Telegraphic Address: "Hallford, London."   Telephone No. 1846.

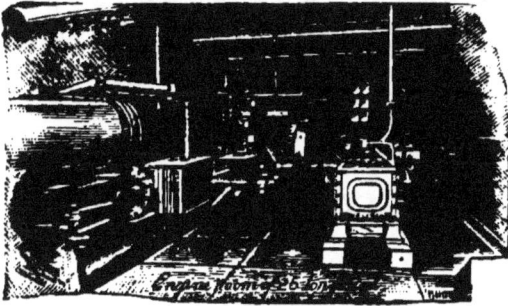

ICE
MAKING
AND
COOLING
PLANT.

FOR MAKING ICE AND COOLING AIR AT THE LOWEST POSSIBLE COST FOR FUEL, LABOUR, AND UPKEEP.

RESULTS GUARANTEED.

REFERENCES TO A LARGE NUMBER OF INSTALLATIONS IN THIS COUNTRY AND ABROAD, GIVING THE MOST SATISFACTORY MECHANICAL AND COMMERCIAL RESULTS.

Write for List Number 1. Post Free.

MAKERS ALSO OF
HAND ICE-MAKING MACHINES.
PUMPS FOR EVERY SERVICE.
ETC. ETC.

THE PULSOMETER

ENGINEERING CO. LD.,
NINE ELMS IRON WORKS,
LONDON, S.W.

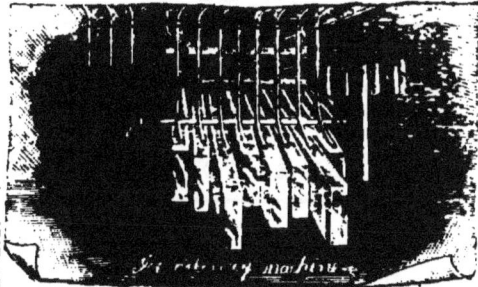

ADVERTISEMENTS.

PETER BROTHERHOOD,

Mechanical Engineer,

15 & 17 BELVEDERE ROAD,

LONDON, E.C.

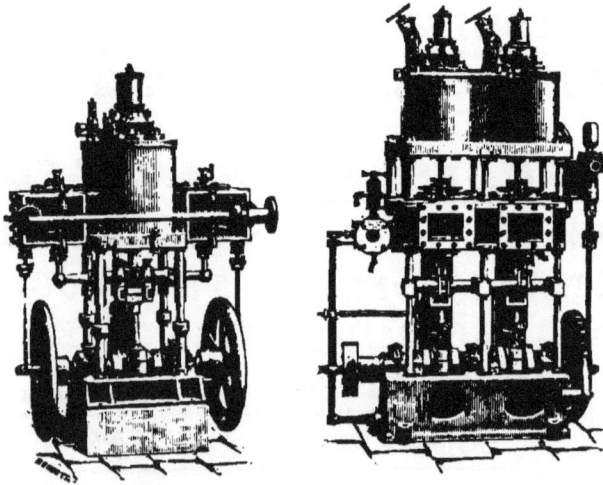

Patent High Pressure Compressor for Air

and Gas, as supplied to Her Majesty's

and Foreign Governments

PARTICULARS AND PRICES ON APPLICATION.

LEASK'S
MARINE ACADEMY,
95 MINORIES,
TOWER HILL, LONDON, E.

ENGINEERS PREPARED FOR

BOARD OF TRADE AND OTHER EXAMINATIONS
(SECONDS, CHIEFS, EXTRAS, & SURVEYORS),
By a New System of Education, which includes

EVERYTHING REQUIRED AT THE EXAMINATION,
And enables Students while solving the most difficult questions to understand the reason why at every step in the process.

This makes all the difference between success and failure.

The Subjects embraced in the ordinary course include — Arithmetic, Drawing, Strength of Materials, Indicator Diagrams, Setting of Slide Valves and Position of Eccentrics, Geared and Oscillating Engines and their Valve Gear, Boiler Construction and Management, Testing of Water-Gauge and Density of Water, Salting, Scaling, Priming, Pitting, Corrosion, Consumption of Fuel, etc. Jet, Compound, Triple and Quadruple Engines and their Management, Breakdowns and how to Repair them. Properties, Expansion and Condensation of Steam, Electricity, and other practical subjects included in the Verbal Examinations.

The Great and Uniform Success of its Students, and the high marks made by some of them after failure elsewhere, clearly proves that no matter how backward any Engineer may be either in Arithmetic or VERBALS, his success may be ensured by studying for a short time at this Academy.

It is a significant fact, proved by the Examination Records, that

More Engineers are Yearly Passed from this Academy than from any other, The best evidence of its remarkable success.

PRINCIPAL TUTORS:
ALEX. R. LEASK, C.E., Consulting Marine Engineer and Surveyor, formerly Engineer Surveyor to H.M. Col. Govt., etc.

A. RITCHIE LEASK, M.E., Certificated First-Class Marine Engineer (Author of "Triple Expansion Engines and Boilers," etc.), Queen's Prizeman in Mechanical Drawing, etc.

Assisted by Competent Tutors in ALGEBRA, GEOMETRY, MECHANICS, MACHINE DESIGN, DRAWING, and ELECTRIC LIGHTING, ETC.

HOURS—10 a.m. to 8 p m.; Saturdays, 10 a.m. to 1 p.m. only.

Before deciding where to go, send for a copy of

☞ **LEASK'S HANDBOOK TO EXAMINATIONS.**
48 Pages. Post Free upon Application.

ADVERTISEMENTS.

DE LA VERGNE
PATENT
REFRIGERATING
AND
ICE MAKING SYSTEM.

THE MOST EFFICIENT, RELIABLE, & ECONOMICAL SYSTEM IN THE MARKET.

Sole Licensees for Great Britain, the Colonies, and British Possessions.

L. STERNE & CO. LᵀᴰD.,

The Crown Iron Works,

GLASGOW.

London Office—28 VICTORIA STREET, WESTMINSTER.

Makers of ICE and REFRIGERATING MACHINERY for the New Skating Rink at Niagara Hall. This is the largest Ice Rink in the World. Many of the most important Installations in this Country and abroad have been made and erected by L. Sterne & Co.

Also makers of small "PONY" Machines suitable for Country Mansions, Hotels, Clubs, Retail Meat and Provision Trades, etc. etc.

CATALOGUES forwarded on application to intending Purchasers. Tenders, and all requisite information freely given.

List of · Publications

OF THE

TOWER PUBLISHING COMPANY, Lᵀᴰ·

95 MINORIES, LONDON, E.

No.	

GENERAL LITERATURE.

Paper Covers.   Price 1/-

1 **A HEROINE OF THE SLUMS,. and other Tales.**

By GEORGE GRIFFITH,

AUTHOR OF "THE ANGEL OF THE REVOLUTION," ETC. ETC.

These Tales form a series of narratives in which are depicted some of the most thrilling situations and startling incidents taken from real everyday life that have ever appeared in print.

"A capital shilling collection of exciting and laughable stories."—*Weekly Times and Echo.*
" A very entertaining shilling's worth."—*N. B. Daily Mail.*
" A collection of cleverly written stories."—*Bristol Mercury.*
" A capital book for a holiday or a railway train."—*Scotsman.*

Paper Covers.   Price 1/-

2 **NEW AMAZONIA: A Foretaste of the Future.**

By MRS. GEORGE CORBETT.

" Pictures with a good deal of cleverness a supposed future state of Ireland —good-humoured satire."—*Scotsman.*
" Mrs. Corbett writes with so much power and sparkle, that she provides an excellent shilling's worth of entertainment."—*Glasgow Herald.*
" I offer my best thanks for the courteous presentation of your work; and I shall cherish the hope that the large and free discussion of social relations, in which you bear a part, may prove beneficial in a world which, undoubtedly, presents ample room for improvement."—*W. E. Gladstone.*

Just Published. Paper Covers. Price 1/-

3 **BLOOD   IS   THICKER   THAN   WATER.**

By GEOFFREY DANYERS. .

A Vision of the re-united Anglo-Saxondom asserting the Dominion of the Seas.

A Political Dream.

" The story is cleverly written, and as its central idea must recommend itself to many readers, the book is sure of success."—*Scotsman.*
" The book is cleverly written, and exhibits the author's close study of the subject."—*Dundee Advertiser.*

No

GENERAL LITERATURE.

Cloth. Price 1/6.

21

PHARISEES UNVEILED.

By Mrs. GEORGE CORBETT,

AUTHOR OF "CASSANDRA," "THE MISSING NOTE," ETC.

"Brightly written. Will help to enliven a dull half-hour very ably."—*Whitehall Review.*

"An eminently readable and capable shilling's worth. A remarkably clever and ingenious story."—*Newcastle Daily Chronicle.*

"A well-written and originally conceived story. The courageous and vivacious style makes the book very readable."—*Leeds Mercury.*

"An ingeniously contrived story. It is entertaining."—*Scotsman.*

Third Edition. In Boards. Price 2/-

41

Mrs. GRUNDY'S VICTIMS. A Realistic Novel.

By Mrs. GEORGE CORBETT.

"The book is clever and the interest is unflagging ; thus the work is likely to have what it deserves—a large circulation."—*Liverpool Mercury.*

"Mrs. Corbett certainly treads upon dangerous ground, but she displays rare taste in the handling of a delicate theme."—*Newcastle Chronicle.*

"Real and horrible evils of our social system are exposed, and terrible defects in the law are revealed."—*Birmingham Gazette.*

"Mrs. Corbett's words leap from page to page in a burning torrent of indictment."—*Morning Leader.*

"A more candid and realistic picture of some of our most glaring social evils we have never met with."—*Weekly Times and Echo.*

"The book is well written, and contains much sorrowful truth on a painful subject."—*Whitehall Review.*

"A very painful story of the evils of slandering tongues and the dangers that beset the path of unprotected innocence."—*Bookman.*

"The moral depravity of the upper classes is exposed in a fearless manner, and in a way that will shock those who know nothing of this phase of existence."—*Dundee Advertiser.*

In Cloth. Price 2/6.

61

Mrs. GRUNDY'S VICTIMS. A Realistic Novel.

By Mrs. GEORGE CORBETT.

No.

GENERAL LITERATURE.

In Cloth. Price 2s. 6d.

62 ## WHEN THE SEA GIVES UP ITS DEAD.

NEW DETECTIVE NOVEL.

By Mrs. GEORGE CORBETT,

AUTHOR OF 'PHARISEES UNVEILED,' 'SECRETS OF A PRIVATE ENQUIRY OFFICE,' 'ADVENTURES OF A LADY DETECTIVE,' 'A SAILOR'S LIFE,' 'THE MYSTERY OF FELLSMERE,' ETC.

"It is only fair to say that the writer is entitled to rank as a very fair second in the *rôle* which Dr. Conan Doyle has created."—*Liverpool Post.*

"Entertaining and well up to the average of this class of story."—*Literary World.*

"The tale is well worth reading." —*Dundee Advertiser.*

"Thrilling and ingenious narrative." —*Admiralty & Horse Guards Gazette.*

"Readable and full of interest."—*Manchester Courier.*

"Mrs. Corbett's book may be safely recommended to all lovers of detective stories."—*Reynolds.*

"It is cleverly written and the interest is well sustained."—*Western Morning News.*

"As a detective story it is equal to the best that have appeared of late."—*Hampshire Telegraph.*

Cloth Gilt. Price 2s. 6d.

63 ## GLORIANA; or, The Revolution of 1900.

By Lady FLORENCE DIXIE,

AUTHOR OF "REDEEMED IN BLOOD," "THE YOUNG CASTAWAYS" "ACROSS PATAGONIA," "ANIWEE; OR, THE WARRIOR QUEEN."

"There is abundant play of fancy in the book, as well as some of the ordinary elements of romance."—*Queen.*

"More interesting than any of our author's previous works."—*Athenæum.*

"A prose Revolt of Islam."—*Saturday Review.*

"It is full of exciting incidents and adventures closely drawn from life." *St. Stephen's Review.*

"The tale is well written, vigorous, and interesting."—*Life.*

No.

GENERAL LITERATURE.

Sixth Edition. Price 6s.

Demy 8vo, handsomely bound in cloth gilt,

101 | # THE CAPTAIN OF THE MARY ROSE,

A TALE OF TO-MORROW.

By W. LAIRD CLOWES,

U.S. Naval Institute,

With 60 Illustrations by the Chevalier de Martino & Fred. T. Jane.

This work has been truly described by the public press as an intensely realistic and stirring romance of the near future. It describes the wonderful adventures of an armour-clad cruiser, built on the Tyne, which takes part in a great Naval War that suddenly breaks out between France and Great Britain. The dashing way in which the vessel is handled, her narrow escapes, the boldness of her successful attacks upon the enemy, and the heroic conduct of her commander and crew, form altogether a narrative of most absorbing interest, and full of exciting scenes and situations.

THE FOLLOWING ARE A FEW PRESS OPINIONS.

" Deserves something more than a mere passing notice."—*The Times.*

" Full of exciting situations. . . . Has manifold attractions for all sorts of readers."—*Army and Navy Gazette.*

" The most notable book of the season."—*The Standard.*

" A clever book. Mr. Clowes is pre-eminent for literary touch and practical knowledge of naval affairs."—*Daily Chronicle.*

" Mr. W. Laird Clowes' exciting story."—*Daily Telegraph.*

" We read ' The Captain of the Mary Rose ' at a sitting."—*The Pall Mall Gazette.*

" Written with no little spirit and imagination. . . . A stirring romance of the future."—*Manchester Guardian.*

" Is of a realistic and exciting character. . . . Designed to show what the naval warfare of the future may be."—*Glasgow Herald.*

" One of the most interesting volumes of the year."—*Liverpool Journal of Commerce.*

" It is well told and magnificently illustrated."—*United Service Magazine.*

" Full of absorbing interest."—*Engineers' Gazette.*

" Is intensely realistic, so much so that after commencing the story every one will be anxious to read to the end."—*Dundee Advertiser.*

" The book is splendidly illustrated."—*Northern Whig.*

No

GENERAL LITERATURE.

Ninth Edition. Price 6s.

Demy 8vo, handsomely bound in cloth gilt,

Uniform with " The Captain of the Mary Rose," with numerous Illustrations by Fred. T. Jane and Edwin S. Hope,

102 | THE ANGEL OF THE REVOLUTION.
A TALE OF THE COMING TERROR.

By GEORGE GRIFFITH.

In this Romance of Love, War, and Revolution, the action takes place ten years hence, and turns upon the solution of the problem of aerial navigation, which enables a vast Secret Society to decide the issue of the coming world-war, for which the great nations of the earth are now preparing. Battles such as have hitherto only been vaguely dreamed of are fought on land and sea and in the air. Aerial navies engage armies and fleets and fortresses, and fight with each other in an unsparing warfare, which has for its prize the empire of the world. Unlike all other essays in prophetic fiction, it deals with the events of to-morrow, and with characters familiar in the eyes of living men. It marks an entirely new departure in fiction, and opens up possibilities which may become stupendous and appalling realities before the present generation of men has passed away.

A FEW PRESS OPINIONS.

" Since the days of the Arabian Nights' Entertainments, we know of no writer who ' takes the cake ' like Mr. George Griffith."—*Daily Chronicle.*

" A really exciting and sensational romance."—*Literary World.*

" As a work of imagination it takes high rank."—*Belfast News Letter.*

" Full of absorbing interest."—*Barrow Herald.*

" This powerful story."—*Liverpool Mercury.*

" An entirely new departure in fiction."—*Reynolds' Newspaper.*

" Of exceptional brilliancy and power."—*Western Figaro.*

" This remarkable story."—*Weekly Times and Echo.*

" There is a facination about his book that few will be able to resist."—*Birmingham Gazette.*

" This exciting romance."—*Licensing World.*

" A work of strong imaginative power."—*Dundee Courier.*

" We must congratulate the author upon the vividness and reality with which he draws his unprecedented pictures."—*Bristol Mercury.*

" Is quite enthralling."—*Glasgow Herald.*

" A striking and facinating novel."—*Hampshire Telegraph.*

No.

GENERAL LITERATURE.

Ninth Edition.   Price 6s.

Demy 8vo, Bound in Illuminated Cloth,
With numerous Illustrations by T. S. C. Crowther & Capt. C. Field.

103 THE GREAT WAR IN ENGLAND IN 1897.

By WILLIAM LE QUEUX, F.R.G.S.,
AUTHOR OF "GUILTY BONDS," "STRANGE TALES OF A NIHILIST," ETC.

This extraordinary and entrancing work is based upon the prognostications of the best living authorities on modern warfare, who have personally assisted the Author. It is the first time an attempt has been made to describe in detail the invasion of Great Britain, and the narrative deals not with the vague, shadowy, and distant, but with the almost immediate future, and in the most graphic manner describes our chaotic condition during the war. Fierce battles—in which all the destructive engines which modern science has devised are brought into play—are fought on land and on sea, and the Author has described them with vivid and appalling realism. Military and naval experts on all hands have pronounced the work absolutely unique, and a most valuable contribution to our literature.

SOME OPINIONS OF GREAT AUTHORITIES.

The DUKE OF CAMBRIDGE has written to Mr. William Le Queux, F.R.G.S. after reading his "Great War in England in 1897," a long letter in which he congratulates the author on the vividness of his interesting forecast, and says—"Such books cannot fail to have a good effect in inducing people to think more seriously of the necessity which lies upon the whole country to always be prepared, and to be more open-handed in giving money for the means of defence."—*Standard.*

LORD ROBERTS, in a long letter to the Author, says—"I have read with considerable interest your vivid account of the dangers to which the loss of our naval supremacy may be expected to expose us. You very properly lay stress on the part which might be taken by the volunteers in the defence of the United Kingdom, and I most thoroughly agree with you as to the value the force might be under such serious circumstances as you depict. Under the conditions specified by you, I should be inclined to regard your forecast of the result of the supposed invasion as being *unduly favourable.* I can only add that I trust such conditions may never arise, and that your estimate of the means immediately available for foreign attacks may be more correct than my own."

LORD WOLSELEY says—"A pleasure to peruse it."

LORD GEO. HAMILTON says it is "striking and original."

SIR CHARLES DILKE says—"I think it is most valuable as tending to make people realize how little we are prepared for war."

Mr. Le Queux has also received letters expressing approbation from the German Emperor, King of Italy, Duke of Connaught Marquis of Salisbury, Lord Alcester, Mr. Gladstone, Mr. J. Chamberlain, and many other distinguished men.

SOME PRESS OPINIONS.

" Mr. Le Queux has special qualifications for the task. He knows a great deal of our Army and Navy, and he is familiar with Continental systems and sentiment. The narrative is lively and spirited, and the author writes with an air of conviction which is calculated to carry the reader on from beginning to end."—*Naval and Military Record.*

" Mr. Le Queux is a vivid writer, and his work gives evidence of care and thoroughness. The chapter dealing with the march of the French on London is particularly fine. The author's production is the best of the kind we have come across for some time. It should emphasise our old contention as to the unreadiness for active service on a prolonged campaign of the sea and land forces of our Empire."—*Admiralty and Horse Guards' Gazette.*

" The story is a capital one, full of interest and incident well sustained and well told."—*Army and Navy Gazette.*

" Everything that can spice a sensational volume."—*The Times.*

" Few works can compare in stirring incidents or careful elaboration of detail with ' The Great War in England in 1897.' . . . A great deal of what he forecasts would be very likely to occur if once England were in the clutches of a strong enemy; and in the manner of description, wherein the tumult and carnage of warfare are brought vividly before the reader. . . A clever and exciting book."—*Morning Post.*

" Mr Le Queux's narrative is well and spiritedly written."—*Sun.*

" The scenes are marked with real and affecting power."—*Birmingham Daily Post.*

" We offer criticism in no carping spirit but as part of our grateful acknowledgement for a brilliant, patriotic, and useful work."—*Sheffield Daily Telegraph.*

" Mr. William Le Queux's volume is certainly the most comprehensive and thrilling of anything yet attempted. Regarded simply as a work of fiction, it is exciting enough to satisfy the most enthusiastic lover of ' blood and thunder' literature. In its more serious aspect—and it is this aspect, of course, which the author desires for it—this book certainly evidences serious thought. . . . It is all very graphic and very thrilling, especially the bombardment of London by the Russians, and the author has not scrupled to avail himself of the latest, even of the future, resources of science "—*Daily Graphic.*

" Mr. Le Queux has handled the strategical and tactical problems with skill and adequate knowledge."—*Evening News.*

" No novel of the day comes up to Mr. Le Queux's "Great War in England in 1897," for excitement. From the preface to the last paragraph he has kept up his prophetic heroics in magnificent style, and if his patriotism does not scatter our indifference to our insular defences, why then nothing will. It is really a terrifying book."—*The Sketch.*

" A very remarkable and important work. There is genius in every line. The descriptions are most realistic, and it is of interest to everybody."—*Il Secolo (Milan).*

" It is not a mere fantastic romance; it is a book to study seriously, and we recommend it to the army and navy of Italy, for it contains many and valuable hints."—*Italia Marinara.*

" The writer's capability to speak regarding his subject is displayed on every page of the book. It is splendidly written."—*Sydney Daily Telegraph.*

Mr. DOUGLAS SLADEN says in the *Literary World:*—" It is undoubtedly one of the books of the year. It is so ingenious and so exciting, it is at once extremely technical and extremely readable. . . . The book is a great book and one that no Englishman could read without a thrill.

No.

GENERAL LITERATURE.

Demy 8vo, handsomely bound in Cloth.   Price 6s.

With Frontispiece by EDWIN S. HOPE.

104 OLGA ROMANOFF; or, the Syren of the Skies.

BY GEORGE GRIFFITH,

AUTHOR OF

" THE ANGEL OF THE REVOLUTION,"   " THE OUTLAWS OF THE AIR."

Dedicated to Mr. HIRAM S. MAXIM.

A sequel to the Author's striking and successful romance, *The Angel of the Revolution*, describing the efforts of a beautiful daughter of the House of Romanoff, to restore the throne to her ancestors destroyed in the World-War of 1904, and presenting to the reader the spectacle of a world transformed into a wonderland of art and science, yet trembling on the brink of a catastrophe, in comparison with which even the tremendous climax of *The Angel* sinks almost into insignificance.

SOME PRESS OPINIONS.

" Mr. George Griffith has made himself a high reputation as an imaginative novelist by his brilliant romances, *The Angel of the Revolution*, and *The Syren of the Skies.*"—*Sketch.*

" This is quite as imaginative, as clever, and as enthralling a book as its predecessor." —*Glasgow Herald.*

" The book is a wild one, but its wildness and imaginative boldness make it uncommonly interesting."— *Scotsman.*

" The flights of fancy and imagination displayed by the author show a most marvellous power and conception."—*Aberdeen Free Press.*

" An entrancing book."—*Birmingham Post.*

" Full of originality in its rendition.   . . .   A marvel of imaginative strength and picturesque pen painting."—*European Mail.*

" On the whole Mr. Griffith has published a work which to our mind is the most suggestive of its kind that has been published for many years."— *Admiralty and Horse Guards Gazet'e.*

" The work hardly lends itself to critical remark other than the expression of one's appreciation of an imaginative and glowing style likely to add to the pleasure of those who enjoy purely speculative fiction.   These pictures have a weird splendour in keeping with the theme, but it is natural to desire a better future for the human race than the one here prophesied."—*Morning Post.*

" His theme is a more tremendous one, and the incidents of his story tenfold more terrible than even those awful battles in the former volume. There is the same swift succession of awful calamities, the same sustained interest from title page to cover, and the same thread of human love running through the narrative which lent its chief charm to the ' Angel of the Revolution ' "— *Weekly Times and Echo.*

' By lovers of sensational writing, in which the scientific discoveries of the future are forecast, and intrigue and warfare related in realistic manner under conditions which now exist but in prophetic imagination, it will be warmly welcomed.   . . .   The book must be read to be appreciated. Description is impossible."—*Bradford Daily Argus.*

No.

GENERAL LITERATURE.

READY SHORTLY.

Demy 8vo, bound in illuminated Cloth. Price 6s.

With Illustrations by HAROLD H. PIFFARD.

105

ZORAIDA,

A ROMANCE OF THE HAREM & THE DESERT,

By WILLIAM LE QUEUX, F.R.G S.,

AUTHOR OF "THE GREAT WAR IN ENGLAND," "GUILTY BONDS,"
"THE TEMPTRESS," ETC. ETC.

READY SHORTLY.

Demy 8vo, handsomely bound in Cloth. Price 6s.

With numerous Illustrations by E. S. HOPE.

106 # THE OUTLAWS OF THE AIR,

By GEORGE GRIFFITH,

AUTHOR OF "THE ANGEL OF THE REVOLUTION," "OLGA ROMANOFF," ETC.

READY SHORTLY. Price 6s.

THE TOWER ROMANCE SERIES.

107 # VOL. 1. A TORQUAY MARRIAGE.

By G. & E. RAYLEIGH-VICARS.

READY SHORTLY.

Vol. II. Price 6s.

108 # IN QUEST OF A NAME,

By MRS. HENRY WYLDE,

AUTHOR OF "SEVERED TIES," "HER OATH," "FATHER AND SON,"
"WRONGED," ETC.

With Illustrations by HAROLD H. PIFFARD.

No.

SCIENTIFIC AND TECHNICAL WORKS.

Eighth Year of Publication. Price 1s.

501 | # The "Engineers' Gazette" Annual and Almanack.

NOW ENLARGED TO 240 PAGES.

PRINCIPAL CONTENTS:—

INTERESTING ARTICLES on Modern Engineering Practice; TIDE-TABLES-for London, and Constants for 368 Ports; TABLES of Areas, Knots, Pressures, Temperatures, &c.; BRITISH Shipbuilders and Safety Valve Makers; ELECTRICAL Engineering Notes, Rules and Formulæ; WHITWORTH Scholarships and Exhibitions for 1894; ENGINEERING Rules, Recipes, Tables and Statistics; DISTANCES and TIME on Ocean Mail Routes; POST-OFFICE, Charges for Letters, Books. &c., Engineers' Department of Postal Telegraphs; NEW SERIES OF NOTES on BOILERS; FOREIGN MONIES, Weights and Measures; BRITISH CONSULATES; ADMIRALTY OFFICIALS, Ships in Royal Navy and their Stations, Reserve Merchant Cruisers; BOARD OF TRADE —Marine Department; LLOYD'S REGISTER—Agents and Surveyors; ENGINEER OFFICERS in R. N. Reserve; READY REFERENCE NOTES FOR ENGINEERS; Ordnance Factories; Boiler Insurance Companies; Navy Estimates for 1894-95, Rates of Pay; Naval Constructors; Factory Inspectors; British Yacht Clubs; and a Miscellany of Entertaining Extracts.

Price 1s. 6d.

521

THE COSMOS; an Essay.

IN TWO PARTS.

Part I.—Hypothesis of Universal Evolution, &c.
„   II.—Phychological and Objective Analysis and Review of Physical Science.

By J. D. SLYTHE.

Price 1s. 6d. Nett.

522 | # WORKING SILDE VALVE MODELS FOR MARINE ENGINEERS,

One complete Set consisting of

No. 1.—COMMON D SLIDE VALVE.
No. 2.—DOUBLE PORTED SLIDE VALVE.
No. 3.—PISTON VALVE (with Steam outside).
No. 4.—PISTON VALVE (with Steam inside).

Being made in Cardboard, with the various parts differently coloured, and so constructed that the Valves may be given any desired travel, while the correct positions of Eccentrics and Crank can be seen at a glance, these Models will be found of very great service in all problems relating to the setting of Valves, and expanding of steam in the Cylinders.

No.

SCIENTIFIC AND TECHNICAL WORKS.

THE PRACTICAL SERIES OF ENGINEERING HANDBOOKS.

Crown 8vo, Cloth Gilt, with 64 Illustrations. Price 2s. 6d.

531 PRACTICAL ADVICE FOR MARINE ENGINEERS.

By CHAS. W. ROBERTS, M. I. Mar. E,

Practical Engineer and Draughtsman.

The purpose of this book is to place together a few practical hints regarding the Management of Marine Engines and Boilers. It should be especially welcome to junior engineers in assisting them to grasp the general ideas which should govern the management of Steamship Machinery.

Crown 8vo, Cloth Gilt, with 134 Illustrations. Price 5s.

541 ELECTRIC LIGHTING FOR MARINE ENGINEERS.

By SYDNEY F. WALKER,

M.I.F.E., M.I.M.E., ASSOC. M.I.C.E., M. AMER. I.E.E., C.E.

This work has been produced to meet the demand by Marine Engineers for a reliable guide to the installation and management of electrical plant on board ship, and is of a thoroughly practical character.

It gives exactly the information required, even to the most minute detail, and contains illustrations of all the approved machines and fittings now in use, with careful instructions as to adjusting and testing them under all conditions.

" Mr. Walker's book is admirably suited for its purpose."—*Industries.*
" The book·will be found to be very suitable for all who want information on these topics."—*Electricity.*
" Is an exceedingly useful little volume. The contents cover a very large ground, and deal very fully with the subject."—*Southampton Times.*
" A veritable boon to hundreds of engineers."—*Engineers' Gazette.*

Crown 8vo, Cloth Gilt, with 59 Illustrations. Price 5s.

542 TRIPLE EXPANSION ENGINES.

AND BOILERS AND THEIR MANAGEMENT.

By A. RITCHIE LEASK,

Author of " Refrigerating Machinery : its Principles and Management," " Breakdowns at Sea, and how to Repair them," &c., &c.

This work is intended to supply Engineers with practical information regarding the latest type of Triple and Quadruple Expansion Engines and Boilers and their management. It is written in a plain and unpretentious style, and embodies the recent experiences of those who have sailed with such Engines and Boilers.

" The work is brimful of practical information of much value to marine engineers."—*Steamship.*
" In it they will find everything that is known, or worth knowing at least, about modern high pressure engines."—*Engineers' Gazette.*
" Cannot fail to be of the very greatest use alike to the student, the engineer, and the general reader."—*Journal of Commerce.*
" The author writes with great clearness and point, and has succeeded in producing a really valuable work."—*Southampton Times.*

No.

SCIENTIFIC AND TECHNICAL WORKS.

THE PRACTICAL SERIES OF ENGINEERING HANDBOOKS—*continued.*

Crown 8vo, Cloth Gilt, with 89 Illustrations. Price 5s.

543 BREAKDOWNS AT SEA, and how to Repair them.

By A. RITCHIE LEASK,

AUTHOR OF "TRIPLE EXPANSION ENGINES AND BOILERS," "REFRIGERATING
MACHINERY, ITS PRINCIPLES AND MANAGEMENT," ETC., ETC.

The great importance of this subject, and the pressing need for some work dealing with it, is sufficient justification for the publication of the above book.

Its contents comprise nearly every conceivable breakdown at sea, and the methods of repair in each case are written in a plain and popular style, so as to be easily followed in similar cases of emergency.

The book is copiously illustrated, and should prove of very great value to all classes of Marine Engineers.

"This is a book for which seagoing engineers will be grateful. The language is simple and direct. The numerous illustrations are all of great practical use."—*Engineers' Gazette.*

"The book will undoubtedly prove of service to engineers, and the hints and suggestions given will aid them in overcoming difficulties arising through the breakdowns of machinery at sea."—*Steamship.*

"It tells us how promptly engineers have faced and overcome danger, and how they have refused to be discouraged by the magnitude of the task set before them, and the inadequacy of their means of combating the difficulties that have arisen in their path."—*Marine Engineer.*

"Handsomely printed and well bound."—*American Shipbuilder.*

Crown 8vo, Cloth Gilt, with 64 Illustrations.   Price 5s.

544 REFRIGERATING MACHINERY, its Principles & Management.

By A. RITCHIE LEASK,

AUTHOR OF "TRIPLE EXPANSION ENGINES AND BOILERS," "BREAKDOWNS
AT SEA, AND HOW TO REPAIR THEM." ETC., ETC.

This work is written in such a style as to be readily understood by all. It goes thoroughly and minutely into the theory of refrigeration, and gives in detail the various methods by which it is accomplished. The management of the machinery is also carefully dealt with, the whole forming a complete and valuable guide to those in charge of refrigerating plant of all kinds.

SCIENTIFIC AND TECHNICAL WORKS.

With 27 Illustrations and 21 Large Plates,

Demy 8vo,· Cloth Gilt.   Price 6s.

561 | # DRAWING & DESIGNING for Marine Engineers.

By CHARLES W. ROBERTS, M. I. Mar. E.,

AUTHOR OF "PRACTICAL ADVICE FOR MARINE ENGINEERS," ETC.

This work has been written to meet the demand for an "up to date" book on drawing and designing marine engines and boilers, which would be suitable for seagoing engineers when preparing for the Board of Trade Examinations, and for those who are unacquainted with the routine of the work carried on in a drawing office.

PRINCIPAL CONTENTS.

Chap. I.—DRAWING MATERIALS AND INSTRUMENTS. Introduction— Materials and Instruments required—The Drawing Board—T and Set Squares—Compasses, &c.—Drawing Pens, Pencils—Paper—Scales—General Instructions—Inking in—Erasures—Alterations—Colouring—Geometry.

Chap. II.—PROJECTION OF DRAWINGS. Vertical, Horizontal and Inclined Plane—Examples of Projection—Angular Projection—Arrangement of Views—Examples of Projection of Curved Lines—Projection of Screws— Shadow Lines—Shop Tracings—Photographic Tracings.

Chap. III.—DESIGNING CYLINDERS AND SLIDE VALVES. Things to be studied when designing—Order of designing the different parts—Horse power required for given speed and size of ship—Type of engine—Diameter of cylinders—Stroke—Revolutions—Drawing the Cylinders—Specification of Cylinders—Slide Valves.

Chap. IV.—PISTON AND CONNECTING RODS, SHAFTING, &c. Piston Rod and Crosshead—Guide Shoe—Guide Plate—Connecting Rod—Crank Shaft— Thrust Shaft—Thrust Block—Propeller Shaft—Stern Tube—Valve Gear— Eccentrics—Bolts—Screws, &c.

Chap. V.—BED PLATE, CONDENSER, PUMPS, PIPES, &c.   Bed Plate— Condenser—Circulating Pump—Air Pump—Feed Pumps—Bilge Pumps— Pump Levers—Centre Bearing—Pipes—Connections—Mud Boxes—Stop Valves—Discharge Valves.

Chap. VI.—THE SCREW PROPELLER. Twin Screws—Slip of Screw— Explanation of Technical Terms—Material of Blades—Designing a Propeller —Moulding a Propeller—Working Drawing—How to project the drawing of a Propeller.

Chap. VII.—MARINE BOILERS. Different forms of Boilers—Materials used—Area of Fire Grate—Heating surface—Rivetting—Strength of Joints—Shell Plates—End Plates—Furnaces—Fire Bars—Combustion Chambers—Tubes—Tube Plates—Stays—Furnace Fittings—Boiler Seatings —Working drawings, &c.

Chap. VIII.—BOILER MOUNTINGS. Stop Valves—Safety Valves—Feed Check Valves—Pressure Guage—Water Guage—Blow-off Cocks—Salmometer Cocks.

No.

SCIENTIFIC AND TECHNICAL WORKS.

Demy 8vo, Cloth Gilt.   Price 12s. 6d.

581 | SOLUTIONS TO QUESTIONS

GIVEN AT THE

Extra First Class Engineers' Examinations

Of the Board of Trade during the last Ten Years.

FULLY WORKED OUT and Illustrated with 147 DIAGRAMS.

By EDWARD J. M. DAVIES, C.E. & M.I.M.E.,

Whitworth Scholar, Millar Prizeman, Inst. C.E ; Honours Medallist of the Science and Art Department, and City and Guilds of London Institute, &c , &c.

· This work has been written specially for Candidates for the " Extra First Class " Certificates granted to Marine Engineers by the Board of Trade, but will be found of great service also to other Engineers who wish to improve by study.

Its contents include all the principal questions given during the last ten years and up to the date of publication.   As each one is worked out in full, and illustrated by diagrams wherever necessary, the possession of this work will enable any Engineer of ordinary ability to master the subjects embraced in this Examination.

The language employed is such as will be readily understood by practical men, and the methods of solution have been simplified to the greatest possible extent, so that they might be intelligible to the average Engineer.

" The anthor is to be congratulated on having produced a book which will be of considerable value to intending candidates.   The explanations are always full, and the formulæ are carefully analysed."—*Steamship.*

" The author may further claim the distinction of having contributed to current literature a work which is in itself unique, constituting a guide to the higher examination, and which is as complete as it is reliable."—*Machinery.*

" The solutions are so cleary expressed and so well illustrated by diagrams, that he must have a very dull brain indeed who cannot readily understand the language used and the sketches provided for his instruction."—*Engineers' Gazette.*

No. | **WORKS SOLD BY TOWER PUBLISHING COMPANY.**

Crown 8vo.   Price 6d.

A **APPLIED MECHANICS.**

PARTS I. and II.

By A. N. SOMERSCALES,

LECTURER ON STEAM, MECHANICS, AND MECHANICAL ENGINEERING AT THE
HULL YOUNG PEOPLE'S CHRISTIAN AND LITERARY INSTITUTE.

This work contains Definitions, Strength of Materials, Laws of Strength,
Useful Numbers, Weight of Materials, and over 400 Questions and Answers
to Problems in Applied Mechanics.

Crown 8vo.   Price 1s.

B **FORMULÆ, RULES, AND QUESTIONS IN STEAM**
For the use of Students in Science Classes.

By A. N. SOMERSCALES.

Containing Useful Numbers, Temperature and Weight of Steam, Mensura-
tion, Weight and Pressure, Historical Engines, the Direct Acting
Engine, the Cylinder and Steam Ports, the Slide Valve and Eccentric,
the Condenser and Air-Pump, the Marine Engine, Expansion of Steam,
Steam Boilers, Measurement of Heat, Thermo-Dynamics, Mechanical
Principles, and Answers to the Numerical Questions.

Crown 8vo.   Price 1s. 6d.   .

C **MECHANICS FOR ENGINEERING STUDENTS,**

Comprising Notes of Thirty Elementary Lessons

On Mensuration, Mechanical Powers, Work, Friction, Horse Power,
Pressure and Buoyancy of Water, Pumps, Gearing, Principle of
Moments, Tension, Shear and Compression, Strength of Pipes, Boilers,
Beams & Shafts, Elasticity of Materials, Centre of Gravity, Centrifugal
Force, Resolution of Forces, Momentum and Energy.

ILLUSTRATED WITH· ONE HUNDRED & FORTY EXAMPLES WORKED OUT,

NUMEROUS DIAGRAMS,

And over Four Hundred Exercises with Answers.

By A. N. SOMERSCALES,

LECTURER ON MECHANICS AND. MECHANICAL ENGINEERING AT THE
HULL YOUNG PEOPLE'S CHRISTIAN AND LITERARY INSTITUTE.

No. | **WORKS SOLD BY TOWER PUBLISHING COMPANY.**

Demy 8vo, 510 pages, with Twelve full-page Plates.   Price 12s. 6d.

A MUCH IMPROVED AND LARGER EDITION OF

D | # CIVIL AND MECHANICAL ENGINEERING

Popularly and Socially considered.

By J. W. C. HALDANE, Consulting Engineer.

"A work of great value, written with conspicuous ability, and rich in instructive and entertaining matter."—*Morning Post.*

"One of the most entertaining as well as instructive books we have ever read."—*Engineers' Gazette.*

"Eminently readable and instructive."—*Western Mail.*

"A most valuable book."—*Shipping Telegraph.*

"Should be in the library of every engineer and machinery user."—*Machinery Market.*

"So far as engineering goes, Mr. Haldane's book is practically unrivalled."—*British Journal of Commerce.*

Demy 8vo, with numerous Illustrations.   Price 15s.

E | # SERVICE CHEMISTRY,

Being a Short Manual of Chemistry and its Applications in the Naval and Military Services.

By VIVIAN B. LEWES, F.I.C., F.C.S,

PROFESSOR OF CHEMISTRY. ROYAL NAVAL COLLEGE, GREENWICH; ASSOCIATE OF THE INSTITUTION OF NAVAL ARCHITECTS, ETC., ETC.

Gratis and Post Free.

LEASK'S ENGINEER'S HANDBOOK

To the Board of Trade Examinations of Engineers

For the Year 1895.

COMPILED BY ALEX. R. LEASK, C.E.,

Consulting Marine Engineer and Surveyor, Inventor of Improvements in Electric Lighting, Formerly Surveyor to H.M. Col. Gov.

Containing Regulations and all necessary information regarding Engineers' Certificates of Competency of the Second, First, and Extra First Classes and Appointments as Engineer Surveyors to Board of Trade, and to Lloyd's Registry of Shipping,

WITH SPECIMENS OF PAPERS GIVEN AT RECENT EXAMINATIONS.

Any Book in this List will be sent Post-free to any address upon receipt of the Published Price.

THE HASLAM FOUNDRY AND ENGINEERING COMPANY L^{D.},

Incorporated with

PONTIFEX & WOOD L^{D.},

UNION FOUNDRY, DERBY, & 34 NEW BRIDGE STREET, LONDON, E.C.

MANUFACTURERS OF

Refrigerating Machines for Cold Storage of Meat,
Frozen and Fresh, and all kinds of Food;

ALSO

For Breweries, Distilleries, Dairies,

and Ice Making

on the

HASLAM COLD AIR, AMMONIA COMPRESSION, AND PONTIFEX ABSORPTION SYSTEMS.

Telegrams—

" ZERO, DERBY."   " PONFEX, LONDON."

REFRIGERATING

AND

ICE MACHINERY

ON THE

LINDE AND LIGHTFOOT SYSTEMS.

OVER 2100 MACHINES SOLD.

ABOUT 200 FITTED ON BOARD SHIP.

PATENT COMPOUND MACHINES
FOR USE IN HOT CLIMATES.

EVAPORATIVE CONDENSERS
FOR A MINIMUM WATER CONSUMPTION.

THE LINDE BRITISH REFRIGERATION
COMPANY LIMITED,

35 QUEEN VICTORIA ST., LONDON, E.C.

TELEPHONE No. 1573.   TELEGRAMS "SEPARATOR, LONDON."

Branch Offices—

LIVERPOOL	.	. CENTRAL BUILDINGS, NORTH JOHN ST.
GLASGOW		. 48 WEST REGENT STREET.
SYDNEY	.	. 97 PITT STREET.
NEW ZEALAND	.	HAWKE'S BAY FOUNDRY, PORT NAPIER.
MONTREAL	.	. IMPERIAL BUILDINGS.

www.ingramcontent.com/pod-product-compliance
Lightning Source LLC
Chambersburg PA
CBHW021508210326
41599CB00012B/1178